生长与蔓延
蔓·设计实践 I

GROWING AND PERMEATED
ORGANICALLY-PERMEATED DESIGN PRACTICE I

王振军　著

中国建筑工业出版社

序

今年六月下旬，王振军郑重地交给我一册厚厚的书稿，是他的新作《生长与蔓延—蔓·设计实践Ⅰ》，并嘱我代为作序。早已见他著述不断，今天又见如此新作令人欣喜。未承想因我繁事不断，作序之事竟拖延至今日。

我与振军相识30余年，1991年他研究生毕业来中国电子工程设计院股份有限公司（以下简称中国电子院）工作之后，我们在繁忙的工作之余，经常讨论的话题都是围绕建筑创作的趣味、工业建筑设计的意义，以及新材料、新做法等。振军对建筑创作的喜爱溢于言表，令人印象深刻。

可能是由于在学校读本科、留校和读研究生的9年经历，振军在建筑创作的同时喜欢进行学术上的思考辨析，爱写文章，常做到建筑设计和理念主张两相融合，从而促发许多新创意。在主流刊物上发表文章30余篇，值得关注的是2017年出版了《蔓·设计》一书，结合30余年设计实践，努力探讨建筑创作的内在规律。同时，大家还在阅读的《轴线手法在当代建筑设计中的应用》一书，是他在校学习期间的研究生课题，历经不断打磨精致成书，内容主要为古典建筑轴线法和当代建筑轴线法的应用系统，同样是文字严谨，图面细密，仔细品读下去，令人爱不释手。在实践上，他一直踏踏实实、兢兢业业地专心在设计一线，不管是工业建筑还是民用建筑，不管是大项目还是小项目，他都认真对待、反复推敲，完成了很多有影响的作品。

2023年王振军工作室+"蔓·设计"研究中心举办了建筑创作15年作品展，展示了设计团队的发展历史和遍及全国的工程业绩，向业界展现出了体系化的"蔓·设计"理论的实践成果，同时邀请了国内著名的院士、大师、总建筑师参加交流会，进行了广泛的学术交流，大家畅所欲言，提出许多宝贵建议，交流会和作品展得到了大家的好评。

王振军工作室的努力，交流会的成功召开，更验证了建筑学发展的正确途径："建筑设计的理论和方法是建筑学的核心，这是因为指导建筑设计创作实践是建筑学的最终目的"❶。

我认为，作为大院的总建筑师，振军所获得的突出成就，首先是重视设计团队的建设和培养，15年来由6名成员开始起步，发展成为30余人的综合设计团队（包括景观、室内、项目的全过程管理）；设计并完成了110余项工程，遍及全国；提出了"蔓·设计"有针对性的理论指引，大家邂逅于此，讨论之后达成共识，少走弯路不跑偏，团队成长为一个大家庭，从容专注地走在创新的路上，避免了理论的焦虑和纠结。

其次是"蔓·设计"理论的创建和形成源于团队的设计实践，这些都是在工作中发现问题引发的思考，而非宏大叙事，是自然的可以落地的理念，经过磨合演变为一个理论课题，因而更有特色和生命力。"蔓·设计"理论把环境和建筑都当作有机体来看待，并由此倡导在建筑设计中向自然学习，学习自然界中的"几何学"，追求有机建构的结果。用"生长与蔓延"这一自然生命现象来感悟建筑创作的过程，观察人的活动，人生活在其中，成为建筑的重要事件。"生长"是哲学意义上开始的自然场地对建筑的锚固，"蔓延"是建构的全过程，在设计中能跟踪到底才是高完成度的保证。

其三，中国电子院是设计先进工业建筑、世界顶级生产环境空间的平台。众多专业人才齐聚，得天独厚的高科技环境，使建筑师具备更加深远的国际视野。设计工业建筑，重要的是了解工艺流程和生产环境要求，在动态观察中完成对建筑空间组合的设想，成为工业建筑设计的切入点。

工业建筑的本质仍是建筑空间，工业建筑把功能性排在第一，反映了建筑空间的真实性。工业建筑的简约特点和建筑形态的陌生感，特别是在这一点上继承了现代建筑的理性内核，同时也展示了高科技工业建筑新的美学特征。

工业设计院背景的建筑师，对大规模的场地和超长、超大的工业厂房有较高的调控能力和较丰富的交叉专业的相关知识，在新技术、新材料运用和新标准实践中，特别是能源节约、环境保护和资源再利用、智能化等方面都做了深入的研究和落实，保证了设计配合的顺畅。

建筑专业的特点常使我们充满工作趣味和拥有建筑创作的情怀。从类型上讲，建筑创作没有止境，建筑师可以完成人们活动需要的各种建筑场景的设计，类型涉猎较广这也是工业院建筑师实践的特点。

本书主要分为思考与言论、设计实践和学术访谈等。

---

❶ 戴念慈，齐康.中国大百科全书，建筑学.北京：中国大百科全书出版社，1988：7.

关于设计实践，在 2017 年出版《蔓·设计》之后又有多项重要项目已经由设计状态进入使用阶段，本次得以补充入书。其中影响较大的有 2010 年上海世博会沙特国家馆，浦东软件园系列，首都机场、北京大兴国际机场配套工程，机场塔台系列，紫光南京集成电路工厂等。经过多年的设计探索，振军和他团队的设计手法愈加娴熟流畅，设计成果丰硕，作品也成为城市的标志性建筑。

"蔓·设计"理论，历经多年的磨砺前行，读来引人入胜，她用自然界生命的真实状态来比喻建筑的过程与场所的锚固，建筑场所要素的显现使建筑的独特性也油然而生。更重要的是人的直接感觉，只有当人体验了场所的意义时它就"定居"了。"生长和蔓延"明确了"如何看待建筑"和"如何表达建筑"，从而使建筑创作有了举重若轻的从容，最终希望的是建筑师的设计应走的是一条充满趣味、轻装前进的路径。

我想上述成果也绝非偶然，振军勤于思考，乐于求索，善于在工程设计中进行理性分析，进而梳理出一些富于哲理的体验。他对建筑创作的思考仍在不断完善、丰富和发展之中，期待"蔓·设计"新作不断问世，预祝他和他的团队累足成步，取得更大的成就。

黄星元
全国工程勘察设计大师
梁思成奖获得者
中国电子工程设计院股份
有限公司顾问总建筑师
2024.8.27

Foreword

In late June this year, Wang Zhenjun solemnly presented me with a thick manuscript, a typeset sample of his latest work titled "Growing and Permeated" (Organically-Permeated Design Practice I), and entrusted me with writing a preface for it. Having long been aware of his prolific writing, I was delighted to see yet another remarkable piece from him. Little did I anticipate that, due to my numerous commitments, the task of writing this preface would be delayed until today.

I have known Zhenjun for over three decades since he joined our institute after graduating with a master's degree in 1991. Amid our hectic work schedules, our frequent discussions often revolved around the joys of architectural creation, the significance of industrial architectural design, and latest materials and techniques. Zhenjun's passion for architectural creation goes beyond words and is truly impressive.

Perhaps due to his nine-year experience of pursuing his undergraduate degree, staying on to teach, and then obtaining his master's degree at the university, Zhenjun enjoys engaging in academic contemplation and analysis alongside his architectural creations. He has a passion for writing and often integrates his architectural designs with his ideological principles, thereby sparking numerous novel ideas. He has published over 30 articles in mainstream journals, with a noteworthy mention being the publication of "*Organically-Permeated · Design*" in 2017, which, based on his more than 30 years of design practice, endeavors to delve into the inherent principles of architectural creation. Additionally, readers are also immersed in his book "*The Application of Axial Approach: in Contemporary Architectural Design*", which originated as his graduate research project during his academic tenure. This work, honed through continuous efforts, presents a system for applying classical and contemporary architectural axis techniques. Its meticulous writing and detailed illustrations make it a compelling read. In practice, he has consistently remained dedicated and meticulous, diligently focusing on the front lines of design. Whether working on industrial or civil architecture projects, regardless of their scale, he approaches each task with seriousness and thorough deliberation, resulting in numerous influential works.

In 2023, WZJ Studio and "Organically-Permeated Design" Research Center held a 15-year architectural creation exhibition, showcasing the historical development of the design team and their project achievements nationwide, presenting the practical outcomes of the systematic "*Organically-Permeated Design*" theory to the industry. Renowned academicians, masters, and chief architects from across China were invited to participate in the conference, engaging in extensive academic exchanges. Everyone freely shared their views, offering numerous valuable suggestions. The symposium, along with the design exhibition, received overwhelmingly positive reviews.

The efforts of WZJ Studio and the successful convening of the conference further validated the correct path of architectural development: "The theory and methodology of architectural design constitute the core of architecture, as guiding the practice of architectural design creation is the ultimate goal of architecture"[1].

As the chief architect of a major institute, Zhenjun's outstanding achievements can be primarily attributed to his focus on building and nurturing his design team. Over the past 15 years, the team has grown from six initial members to a comprehensive design team of over 30 individuals, encompassing landscape, interior, and full-process project management. They have designed and completed over 110 built projects, with their footprint extending across the entire country. Zhenjun proposed the targeted "*Organically-Permeated Design*" theory, which served as a guiding framework. This theory has enabled team members to collaborate effectively, reach consensus through discussions, avoiding unnecessary detours, and keeping them on the right path. The team has grown into a close-knit family, confidently navigating the path of innovation, free from theoretical anxiety and confusion.

Additionally, the creation and formation of the "*Organically-Permeated Design*" theory stems from the team's design practice. It is sparked by issues encountered during our

---

[1] Dai Nianci, Qi Kang. "Encyclopedia of China, Architecture". Beijing: Encyclopedia of China Publishing House, 1988:7.

work rather than grand narratives. It is a natural and implementable concept that evolved into a theoretical topic through practice and refinement, thereby gaining distinctiveness and vitality. The "*Organically-Permeated Design*" theory treats both the environment and architecture as organic entities, advocating for learning from nature in architectural design. This includes studying the "geometry" found in nature and striving for organically constructed outcomes. The theory draws inspiration from the natural phenomena of "Growing and Permeated" to guide the architectural creation process, observing human activities and daily life, making them key events in architecture. "Growing" symbolizes the building's philosophical anchoring to its natural site, while "Permeated" encompasses the entire construction process. Continuous design follow-up ensures a high level of completion and quality.

Thirdly, the China Electronics Engineering Design Institute serves as a platform for designing advanced industrial buildings and world-class production environments. This institute attracts numerous professionals, and the unique high-tech environment fosters architects with a profound international perspective. When designing industrial buildings, it is crucial to understand the production processes and the requirements of the production environment. The envisioning of architectural space combinations is achieved through dynamic observation, making it a critical entry point for industrial building design.

Despite their specific functions, industrial buildings remain fundamentally architectural spaces. The primary focus on functionality reflects the authenticity of the architectural space. The simplicity of industrial buildings, along with the unfamiliarity of their architectural forms, particularly inherit the rational core of modern architecture. Simultaneously, they also showcase the new aesthetic characteristics of hi-tech industrial buildings.

Architects with a background from industrial design institutes possess advanced control and regulation capabilities for handling industrial plants characterized by their grand scale, extensive dimensions, and oversized volumes. Furthermore, they possess a wealth of interdisciplinary knowledge. In-depth research and implementation are essential in the areas of new technologies, material applications, and standard practices - particularly in energy conservation, environmental protection, resource reuse, and intelligent systems to ensure seamless design coordination.

The characteristic of architectural profession often fills us with the joy of work and the passion for architectural creation. In terms of variety, architectural creation knows no bounds, allowing architects to create diverse built environment to satisfy various needs for human activities. The broad range of building types they engage with is a characteristic of the practical outcomes of architects from industrial design institutes.

The book is organized into three main sections: theoretical articles, architectural design practice, and academic interviews. Regarding the architectural design practice, several significant projects have transitioned from the design phase to the usage phase since the publication of "*Organically-Permeated Design*" in 2017, and are included in this book. Notable projects include the Saudi Arabian Pavilion at the 2010 Shanghai World Expo, the Pudong Software Park series, supporting projects for the Capital Airport and Beijing Daxing International Airport, airport control towers, Tsinghua Unigroup Nanjing Integrated Circuit Base Project, and so on. After years of design exploration, the design techniques of Zhenjun and his team have become increasingly sophisticated and fluid, resulting in a wealth of successful outcomes. These works have also become iconic landmarks in their respective cities.

The "*Organically-Permeated Design*" theory, after years of refinement, is fascinating to explore. It uses the authentic state of life in nature as a metaphor for the process of architecture and the anchoring of places. The manifestation of the site elements of architecture naturally gives rise to the building's uniqueness. What matters most is direct human perception. Only when people experience the significance of a place does it become "settled". "Growing and Permeated" clarifies "how to view architecture" and "how to express architecture", thereby instilling a sense of ease and confidence in architectural creation. Ultimately, the hope is that architects will embark on a path forward with lightness and joy.

The aforementioned achievements are far from coincidental. Zhenjun's diligent thinking, eagerness to explore, and aptitude for rational analysis in engineering design have enabled him to derive philosophically rich experiences. His reflections on architectural creation continues to evolve, expand, and develop. We eagerly anticipate new works under the "*Organically-Permeated Design*" theory and wish him and his team continued success and even greater accomplishments in their future endeavors.

Huang Xingyuan
National Master of Engineering Design and Geotechnique Investion,
Liang Sicheng Award Winner,
Consulting Chief Architect of China Electronics Engineering Design Institute Co., Ltd.
August 27, 2024

自 序

2023年是中国电子院（CEEDI）成立70周年，2024年是王振军工作室（以下简称工作室）成立15周年，这无疑是一个非常关键的节点，在这个重要的历史时刻，我们需要一个回顾、见证、总结和展望的机会。

15年，在历史长河中，犹如白驹过隙，转瞬即逝，而对于工作室而言，则意味着5000多个日日与夜夜，分秒可见。工作室从成立时6个人的建筑创作团队，发展成目前近30人，涵盖策划、建筑设计、景观设计、室内设计以及产品设计的综合设计中心；从一张张草图到110余项大大小小的建成项目遍布全国；更重要的是，我们从繁重、片段性的创作中摸索出了体系化的"蔓·设计"实践之路。15年来，工作室之所以能够一直稳步发展，得益于这一理论的指引，使我们的创作得以应对众多项目所带来的复杂、多元的挑战。

一位前辈说得好："搞艺术就是两件事，一是怎样看待这件事，另一个就是如何表达"。"蔓·设计"理论正是回答了"怎样看待建筑"以及"如何表达建筑"这两个核心问题。

我在2017年出版的《蔓·设计》一书中提出了"蔓·设计"理念，之后得到业内的广泛关注，在社会上也有一些积极的反响，当时主要是基于自己从业30年以来的设计思考做的一种观点总结，希望找到设计实践内在的规律和共性。之后7年来，伴随着大量的理论学习、设计实践、学术交流，特别是通过工作室15年作品展，向国内院士、大师和总建筑师们面对面的请教，受益匪浅、收获颇丰，在此基础上，我对该理念的表述框架做了更系统的梳理，并沿着该理念的两个维度"生长与蔓延"及其交集——"有机建构"做了进一步的展开阐述，同时提炼出一系列的原则和策略，希望能为团队的实践提供更为体系化的指导。

"蔓·设计"理论框架的确立也使我们对"好建筑"有了更清晰的认识：良好的在地性、形式与内容统一、空间体验感人、具有可持续性、关照文化传承、较高的感知度、突出的创新性等。建筑创作靠团队，随着团队对"什么是好建筑"达成共识，工作室无论是在进行创作时的方案构思、方向和手法选择，还是在后期对各个专业综合协调过程中，包括应对外部干扰时，都不再纠结，在创作思路上更加富有逻辑，在设计表达上更加自信，创作过程也更加顺畅，团队创作的作品也得到越来越多业内外专家、朋友的认可，随着作品品质的不断提升，相应地市场也呈现出一种良性状态，这都为工作室的持续创新奠定了坚实的基础。那接下来在"手法可能性的探索""新切入点的寻找"，特别是"艺术想象力的激发"等方面，就成为我们未来挖掘创新潜能的关键，这也让我们对工作室的未来充满期待。

本书以"生长与蔓延"为题，也是希望工作室在项目类型、地域分布以及作品数量等方面不断扩展，更是希望随着"蔓·设计"团队人员的培养和与中国电子院内的不断交流、输送，使这一理念得到进一步传承、延续、光大。

感恩CEEDI，是她给了我们这个能够与世界最先进工程技术、先进材料和新标准无缝衔接的平台；是她让我们这些建筑师对建筑的类型没有了成见，从而有机会在更多样、综合难度更大的建筑类型中得到了历练，特别是应对尺度超大、工艺复杂精密、智能化水平先进的高科技工业建筑的能力获得了提高，这一特点也会在本书中有所呈现。基于所处的信息时代和CEEDI的业务背景，我们提出了"蔓·设计"指导下的、突破传统"民用和工业二元分类法"的、提倡超越内容和类型的"大建筑观"——凡是满足人类活动需要的物理和精神空间的建构皆为建筑，旨在为响应国家从求量转向求质的发展战略，倡导建筑师对这些新型建筑类型以及工业建筑的创作投入更多关注、更多研究。

正如曾经对中国思想界和教育界有重大影响的美国哲学家约翰·杜威（John Dewey）所说："人们说哲学始于惊奇，终于理解。艺术是从已被理解的事物出发，最后结束于惊奇。"我们对建筑设计的认知还有很长的路要走，毫无疑问，思考和实践将是建筑师永恒的工作主题。思考——就是要探究建筑的真谛，从而更深刻地理解建筑的本质；实践——就是在研究和理解建筑的基础上大胆地、不懈地去创作新作品。

本书收录了工作室成立前后、主要是这15年来的论文、学术访谈和主要设计作品，以此作为对过往思考和实践的总结，也是继往开来的起点。建筑创作是一个团队的远行，保持建筑的初心、坚守"蔓·设计"的理念，我们在路上……

王振军
2024.5.30
于北京五路居

Preface

The year 2023 marked the 70th anniversary of the founding of China Electronics Engineering Design Institute (CEEDI), and in 2024, we celebrate the 15th anniversary of our studio's inception. Undoubtedly, this represents a crucial juncture. At this significant historical moment, we require an opportunity to review, witness, summarize, and look forward.

Fifteen years, passed swiftly like fleeting clouds in the vast expanse of history; yet for our studio, it represents more than 5000 days and nights of relentless endeavor, visible in every minute and second. From its original six-person architectural creation team, the studio has evolved into a nearly 30-member comprehensive design center encompassing planning, architectural design, landscape, interior, and product design. From initial sketches to over 110 projects of varying scales completed nationwide, what is more critical is our exploration and establishment of a systematic "Organically-Permeated Design" practice amidst intensive and fragmented creative processes. The studio's steady development over the past 15 years can indeed be attributed to the guidance provided by this theoretical framework, enabling us to cope with the complexity and diversity of challenges presented by numerous projects.

An old-timer veteran once aptly stated, "Art revolves around two things: how we perceive it and how we express it". The "Organically-Permeated Design" theory precisely addresses these two core questions: "how to perceive architecture" and "how to express architecture".

In my 2017 publication, "*Organically-Permeated Design*", I introduced the "Organically-Permeated Design" concept, which subsequently garnered widespread attention from the industry insiders and positive societal feedback. At that time, it was primarily a summary of design reflections based on my 30 years of professional experience, aimed at uncovering the intrinsic rules and commonalities in design practice. Over the following seven years, enriched by extensive theoretical study, design practice, and academic exchanges—particularly through face-to-face consultations with domestic academicians, masters, and chief architects at the studio's 15-year exhibition of works—the benefits and insights have been profound. On this foundation, I have systematically reorganized the conceptual framework of the theory, further elaborating on its two dimensions, "growing and permeated" and their intersection—"organic tectonics." In doing so, I have distilled a series of principles and strategies, hoping to provide more systematic guidance for our team's practice.

The establishment of the "*Organically-Permeated Design*" theoretical framework has also led us to a clearer understanding of "good architecture": strong localization, unity of form and content, emotionally engaging spatial experiences, sustainability, caring for cultural heritage, high perceptibility, and remarkable innovation, among others. Architectural creation relies on teamwork. With the team's consensus on "what constitutes good architecture," whether in the ideation and selection of approaches during the creative process, the coordination among various specializations in later stages, or dealing with external disruptions, the studio no longer hesitates. Our creative approaches are more logical, our design expressions more confident, and our creative processes smoother. The works created by our team have gained increasing recognition from experts and friends both within and outside the industry. As the quality of our work steadily improves, the market likewise reflects a healthy, positive dynamic, all of which lay a solid foundation for the studio's sustained innovation. Moving forward, "exploration of methodological possibilities", "seeking new entry points," and especially "stimulating artistic imagination" will become key to unlocking our innovative potential, and also filling us with anticipation for the studio's future.

This book, titled "Growing and Permeated," not only aims for the studio to expand in project types, geographical distribution, and the number of works but also hopes to perpetuate and magnify the "Organically-Permeated Design" philosophy through the training and continuous exchange and contribution of personnel within our institute.

We are grateful to CEEDI for providing us with a platform that seamlessly connects with the world's most advanced engineering technologies, materials, and new standards. It has allowed us, architects to shed our preconceptions about architectural types, thereby gaining experience in a more diverse array of architectural genres with greater comprehensive challenges, especially in terms of capabilities in handling large-scale, complex precision, and advanced intelligent high-tech industrial buildings, a feature that will be presented in this book. Based on the information age and CEEDI's business background, we propose the "Macro Vision of Architecture" under the guidance of "Organically-Permeated Design", which breaks through the traditional "civilian and industrial binary classification", advocating for a focus beyond content and type—any physical and spiritual space that meets human activity needs is considered architecture. This aims to respond to the country's development strategy shifting from quantity to quality, calling for architects to pay more attention and conduct more research on these new types of buildings and industrial buildings.

As John Dewey, the American philosopher who significantly influenced China's intellectual and educational realms, once said, "Philosophy begins in wonder and ends in understanding. Art starts with what is understood and ends in wonder." Our understanding of architectural art has a long journey ahead, and undoubtedly, contemplation and practice will remain the eternal themes of architects' work. Contemplation means exploring the truth of architecture to gain a deeper understanding of its essence; practice involves boldly and tirelessly creating new works based on the study and understanding of architecture.

This book compiles papers, academic interviews, and major design works from before and after the studio's establishment, particularly over the past 15 years, serving as a summary of past reflections and practices and as a starting point for future progress. Architectural creation is a team's journey. By staying true to the cause of architecture and adhering to the "Organically-Permeated Design" philosophy, we continue on our path…

Wang Zhenjun
May 30, 2024
in Wuluju, Beijing

目 录

序 / 黄星元 / ii
自序 / 王振军 / vi

"蔓·设计"在主张什么 / 2

学术思考
思维结构的调整与建筑文化的新图景 / 24
精神与形式——关于安藤忠雄与后现代主义 / 32
轴线手法与建筑空间及形式的层次感塑造 / 40
建筑的本质与中国建筑师的文化使命 / 51
质疑的时代更需要标准 / 54
像做博物馆一样做工业建筑 / 57
高科技园区规划设计和建设的思考 / 63

设计实践
已建项目
上海浦东软件园一期、二期工程 / 72
首都国际机场东区塔台 / 80
北京国际财源中心（IFC） / 86
上海国家软件出口基地（浦东软件园三期） / 92
2010年上海世博会沙特国家馆 / 100
  建筑秀场上的文化容器——沙特馆 / 110
西安咸阳国际机场新塔台及附属建筑 / 112
长沙中电软件园总部大楼 / 116
固安规划展馆 / 124
北京泰德制药股份有限公司研发中心扩建 / 128
中国信达（合肥）灾备及后援基地 / 132
  庭·源——中国信达（合肥）灾备及后援中心 / 143
郑州新郑国际机场二期扩建空管工程塔台小区土建及配套工程 / 144
  塔的故事——郑州新郑国际机场塔台及航管楼设计 / 150
中央美术学院燕郊校区教学楼 / 152
燕郊世界华人收藏博物馆 / 156
海盐杭州湾智能制造创新中心 / 160
  园区空间的多义性表达——海盐杭州湾智能制造创新中心创作体验 / 172
北京中关村移动智能创新园 / 174

潍坊崇文新农村综合服务基地 / 178
北京大兴国际机场信息中心（ITC）、指挥中心（AOC）/ 182
北京大兴国际机场生活服务设施工程 / 192
  阳光之城——北京大兴国际机场生活服务设施工程设计 / 200
北京大兴国际机场工作区车辆维修中心工程 / 202
中国残联北京按摩医院扩建项目 / 208
  形式与精神——中国残联北京按摩医院创作体验 / 218
北京大兴国际机场非主基地航空公司生活服务设施工程 / 222
烟台经济技术开发区景观卫生间项目 / 228
烟台蓬莱国际机场信息中心 / 232
烟台工程职业技术学院南校区工程 / 236
烟台中科先进材料与绿色制造山东省实验室 / 240
烟台光电传感产业园项目 / 244
北京大兴国际机场行政综合业务楼 / 248

在建项目
紫光南京集成电路基地项目（一期）/ 252
潍坊云创金谷 / 260
珠海机场新建塔台及配套项目 / 264
厦门天马光电子有限公司第 8.6 代新型显示面板生产线项目 / 268
广州大湾区数字经济和生命科学产业园项目 / 272

未建项目
中国国学中心竞赛 / 276
中国瑞达投资发展集团公司瑞达石景山路 23 号院科研基地 / 280
中国计算机博物馆项目概念性规划方案设计 / 284

学术访谈
工作室十周年之际访谈王振军 / 290
我的建筑认知体系：蔓·设计 / 296
AT 灵魂三问"蔓·设计"与"大建筑观" / 302
"蔓·设计"中的蔓访谈 / 304
王振军：生长与蔓延——《当代建筑》CA / 311

关于我们
重要作品获奖 / 316
王振军工作室 15 年展活动纪实 / 320
产、学、研及生活实录 / 324

参考文献 / 332
后　记 / 333
大事记 / 337

Contents

Foreword / Huang Xingyuan iv
Preface / Wang Zhenjun viii

The Propositions of "Organically-Permeated Design" / 13

Academic Reflection
The Adjustment of Thinking Structures and New Perspective of Architectural Culture / 28
Spirit and Form: Tadao Ando and Postmodernism / 36
Axial approach and hierarchical shaping of architectural space and form / 46
The Essence of Architecture and the Cultural Mission of Chinese Architects / 52
The Necessity of Standard in Oppugning Time / 55
Design Industrial Buildings As Museums / 60
The Reflections on the Planning, Design, and Construction of Hi-Tech Parks / 66

Design Practices
Completed Projects
Pudong Software Park, Phase I, II In Shanghai / 72
Control Tower of Beijing Capital International Airport East District / 80
Beijing International Finance Center / 86
Shanghai National Software Export Base (Pudong Software Park Phase III) / 92
The Saudi Arabia Pavilion at Shanghai Expo 2010 / 100
Xi'an Xianyang International Airport New Tower and Annex Buildings / 112
The Headquarters Building of Changsha CEC Software Park / 116
Gu 'An Urban Planning Exhibition Hall / 124
R&D Center Expansion Project of Beijing Tide Pharmaceutical Co., Ltd. / 128
China Cinda (Hefei) Disaster Recovery and Back-Up Base / 132
Control Tower and Supporting Buildings of Zhengzhou Xinzheng International Airport Expansion Project Phase II / 144
Teaching Building of Central Academy of Fine Arts Yanjiao Campus / 152
Yanjiao Collection Museum of Worldwide Chinese / 156
Haiyan Intelligent Manufacturing Innovation Center / 160
Beijing Zhongguancun Mobile Intelligent Innovation Park / 174
Weifang Chongwen New Rural Comprehensive Service Base / 178
The ITC and AOC of Beijing Daxing International Airport / 182
Beijing Daxing International Airport Life Service Facilities Project / 192
Beijing Daxing International Airport Work Area Vehicle Maintenance Center Project / 202
China Disabled Person's Federation Beijing Massage Hospital Expansion Project / 208
Beijing Daxing International Airport Non-Main Base Airline Living Service Facilities Project / 222

Yantai Economic and Technological Development Zone Landscape Toilet Project / 228
Yantai Penglai International Airport Information Center / 232
South Campus Project of Yantai Engineering & Technology College / 236
Yantai Zhongke Advanced Materials and Green Manufacturing Shandong Provincial Laboratory / 240
Yantai Photoelectric Sensing Industrial Park Project / 244
Beijing Daxing International Airport Administrative Comprehensive Business Building / 248

Projects under Construction
Tsinghua Unigroup Nanjing Integrated Circuit Base Project Phase I / 252
Weifang Yunchuang Golden Valley / 260
Zhuhai Airport New Air Traffic Control Tower and Supporting Building Project / 264
Xiamen Tianma Optoelectronics Co., Ltd. 8.6 Generation New Display Panel Production Line Project / 268
Guangzhou Greater Bay Area Digital Economy and Life Science Industrial Park Project / 272

Unbuilt Projects
China Sinology Center Competition / 276
China Ruida Investment and Development Group Corporation Ruida Shijingshan Road No. 23 Academy Scientific Research Base / 280
Conceptual Planning Scheme Design of The Computer Museum Project / 284

About Us
Awards / 318
WZJ Studio 15-Year Exhibition Documentary / 320
Industry, Academia Research and Life Record / 324

Epilogue / 334

"蔓·设计"在主张什么

The Propositions of "Organically-Permeated Design"

"蔓·设计"在主张什么

王振军

引言

"蔓·设计"是本人基于对从事建筑设计30多年的思考和实践进行归纳、提炼和总结，并在工作室进行一系列的实践检验后，于2017年提出的设计理念。工作室在这一共识的引领下，开展了富有成效的创作活动，完成了诸多作品，同时对外通过展览和学术交流，对内通过理论学习并结合项目中遇到的实际问题进行复盘、讨论，我们不断地对该理念进行了补充、修正和完善，而且这一工作还会一直持续下去。

七年之间我们一起经历了令人刻骨铭心的突发公共卫生事件，为此工作室很多项目拖后或取消，但这并不重要，重要的是现实更加促使建筑师去认真审视人与自然、建筑与自然的关系，以及什么是有机建筑。"有机建筑"的确不是一个新概念，从弗兰克·劳埃德·赖特（Frank Lloyd Wright）到勒·柯布西耶（Le Corbusier）、安东尼奥·高迪（Antonio Gaudi）、霍华德（Howard）、阿尔瓦罗·西扎（Alvaro Siza）、丹下健三、贝聿铭、圣地亚哥·卡拉特拉瓦（Santiago Calatrava）、伊东丰雄、RCR、丹·皮尔森（Dan Pearson）等前辈、大师们都一直不懈地在实践和理论上进行着探讨，这正是因为自然本身所具有的层次复杂、形式多元、系统庞大等特点以及"自然"这个概念对建筑师来说的特殊意义——她是建筑师创作的基底所在，如何看待和处理与她的关系是建筑师必须全程直面的首要课题。

本文旨在把对上述问题进行探讨的过程中，我们对建筑学的基本构成进行的再认识，在哲学和方法论层面对"怎样看待以及如何表达建筑"进行的思考和实践，向大家做一个阶段性的汇报。

一、缘起——实践和理论的焦虑

一）缘起一——实践需要

1. 让自己想清楚

2016年正好是我本人本科毕业30年。很长一段时间，伴随着设计实践的进行，关于建筑学的理论焦虑一直困扰着我。因此，在这个节点很有必要对自己所从事专业的认知系统进行一下梳理、总结，对建筑学的本体做一个完整的、系统性的厘清。面对当代建筑理论的多元化与复杂性，以及业界在一些概念上的众说纷纭，我们的确需要找出建筑学的真相，并希望以某种方式在实践中对其进行检验。

2. 向年轻人讲明白

工作室成立后，特别是在2013-2015年间，随着业务的拓展，工作室由6个人扩招到了近20人，面对不断加入的年轻人，时时需要我清晰明了地告诉他们建筑到底该怎么做？什么是"好建筑"？如何在复杂的要素中抓住重点？对此，团队急需达成共识。在此过程中，我也在尝试去实现自己留校时想突破的老观念——"建筑学只能学、无法教"的愿望。

3. 一个团队为什么需要有一个明确的基本共识？

（1）坚守认知：互联网时代信息的泛滥导致知道得越多，失去得越多，最后造成自我迷失，故团队成员需要明确认识，保持辨识能力。

（2）提高效率：建筑创作是团队工作，达成共识，可以让团队成员少走弯路、不纠结、不跑偏，从而促进效率提升。

（3）应对来自甲方、管理部门等外部的压力和干扰：随着甲方的见识越来越广，他们表现得越来越自信而强势，建筑师特别是年轻建筑师如果没有坚定的职业原则，极易变为甲方各种风格的"印钞机"。

二）缘起二——学术思考

1. 人与建筑、场地、环境再到自然的关系——是建筑师需要长期思考，同时也是必须时刻面对、回答的问题。

人类总自诩为世界上最完美的结构，也总苛求世界上的一切事物都完美起来，尤其在今天，人类的能力已无法控制地膨胀起来，使这种追求完美的欲望变得更为强烈。建筑作为人类活动的一种产物，充分地反映了人类的这种活动规律，建筑和城市这种人工物态正是人类在不断追求和探索完美中发展起来的。不幸的是，建筑和城市在显示人类力量和伟大的同时，所呈现的结果却并不令人满意，建筑这一人工物态和自然环境越来越呈现出矛盾和对抗，而且这种对抗还在不断加剧。自然创造了人类，人类又创造了建筑和现代化的城市以及现代生活。但现代生活使我们越来越远离了自然，并导致对自然的无数曲解，其结果是使我们正在失去机会来把握这一能量的本质。

2. 人类历史在西方被看作为人类与自然斗争从而征服自然的历史，而事实证明，这种认知是有问题的。正是这样的对立给我们带来很多麻烦，使我们忽视了生活与环境的密切关系，也使我们自以为可以超越自然法则，可以不受自然世界的惩罚。突发公共卫生事件的产生和结束不得不使我们再次陷入深思。

3. 由于当代建筑理论的多元性与复杂性——原有现代主义建筑理论加上新理论的转码，呈现出百家争鸣，同时也莫衷一是的状态。泛滥的信息使建筑师正在经受考验，建筑师需要一颗"定心丸"，急需有自己的哲学来坚守，以防在概念上的纠缠不清。否则，将极易导致以"他人之新为新""鹦鹉学舌"的结果。

4. 实践证明没有理念的建筑师，容易陷入认知的泥潭而纠缠不清，从而导致方法的无序和凌乱，最后注定也很难走远。日本建筑强调传承的同时鼓励思辨，其建筑师各具特点的设计理念和风格至今在世界建筑圈内获得赞誉，迄今他们在普利兹克建筑奖中获奖人数占到六分之一，这就是明证。

二、建筑学的要素构成梳理

建筑学涉及要素包罗万象，探究建筑的核心问题，就是要对建筑学所涉及的这些构成要素进行分类、归纳，分出主次和先后。透过现象看本质，通过分析，我们将建筑学的构成要素分成了三个层次（图1）：

一）建筑的自治性

包括场地、空间、形体、结构、材料、比例、色彩等，它可以内部立法，不依赖于领域之外的其他因素，是一个学科赖以建立的核心要素。一个学科没有内在性，一切无从谈起。任何专业应先有核心，然后才会有边界可言。

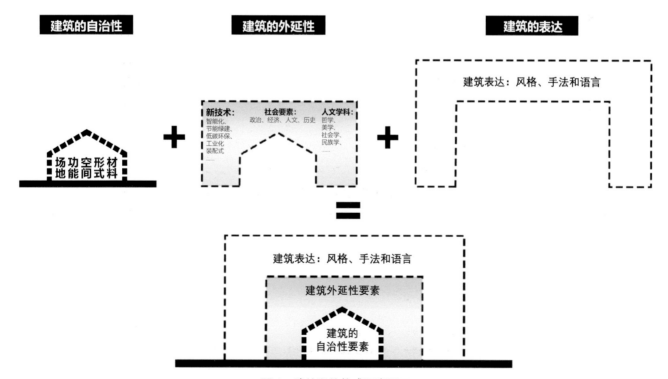

图1 建筑学的构成要素图示

建筑学构成要素的三个层次划分，这三个层次之间的边界是在变化和转化之中的，故这三者之间用虚线表示。

二）建筑的外延性

这包括：

1. 新材料、新技术，或者是侧重于新的技术、材料和社会环境问题所引起的建造和使用方式的变化。

2. 社会要素——政治、经济、文化、人口。

3. 人文学科——哲学、环境学、社会学、人类学等，借助于其他学科的发展来展开有关建筑理论的研究。

三）建筑的表达

涉及风格、手法、建筑语言等。关于建筑的表达，例如手法的选择，针对自然场力中的同一要素，建筑师可以用保存、反射、过滤、隔绝和稀释等各种态度和方法来为建筑的建构服务。要说清楚这个问题，可以拿扎哈·哈迪德（Zaha Hadid）和大卫·奇普菲尔德（David Chipperfield）来举例，两人为英国伦敦建筑联盟学院（AA）同学，都成绩斐然，并且都试图在设计时寻找与城市和左邻右舍的关系，理念似乎相通，但由于两人表达建筑的手法大相径庭，呈现出来的风格也截然不同。

本人认为，在中国当代语境下来讨论、梳理建筑学构成要素及其层次，除了上述提到的因素外，还具有特殊的、紧迫的现实需求和意义，这是因为：一方面现代建筑学的传统和基本内核在中国尚未完全真正确立，建筑学外延因素对于建筑创作的影响在中国也就更为突出。另一方面，随着中国的改革开放，国外建筑师的创作在中国拥有了前所未有的实现机会，当代中国建筑已经卷入国际建筑语境之中。在这种情况下，中国建筑师如何保持清醒，回到建筑本身，在新的视野下来审视这些主题，审视这些要素在当代建筑学中的定位和转化，更是成为当今建筑学必须面对的一个课题。

三、为什么强调向自然学习

一）现场踏勘时，当我们俯瞰基地或俯身仔细观察一砖一石、一草一木，不难发现其中充满了自然力的和谐统一。她们都清晰地显示出来了她们的秩序，这些秩序被精心地组织和规划，展示出的每个细节都充满了建构之美，其中浑然天成的秩序之美、形式与内容统一的和谐之美，令人迷恋和感动。这种现象被雅克·赫尔佐格（Jacques Herzog）称作"自然隐匿的几何学"，他说："存在于自然中的关系系统的复杂性，具有一种不可清楚探究的完美性，我们感兴趣的是这种关系在艺术和社会领域的表现，因此我们关注自然隐匿的几何，它主要是一种精神上的准则，而不是自然外在的表现"❶（图2）。

德国诗人诺瓦里斯（Novalis）说："艺术的秘密就是要使每一个自然现象、自然秩序转换成一种规则——类似建造艺术"。20世纪70年代，凯文·罗奇（Kevin Roche）在被记者问到会如何设计IBM博物馆时，他回答道："我们要建立对高技术及其复杂性的一种内在展示，但这种展示是以某种方式与自然的环境联系在一起，自然才具有最终的复杂性"❷。

二）当下再次强调向自然学习，是基于以下几点原因：

1. 自然界的平衡法则和生态系统的运作方式是人类理解和应用科学原理的基础，对自然的感知和体验是人类形成认知体系的基础。

2. 自然一直是人类灵感的源泉，自然界的多样性和复杂性的确激发了人类的创造力，特别是在艺术创作方面，当然也包括建筑师。

3. 仍在由人类制造中的丑陋建筑，与令人迷恋的自然秩序之美形成强烈的对比，证明人类的成果和能力并没有超越自然，所以我们应向自然学习。

4. 西方生态伦理思想的形成过程给我们的启发：

这一理论形成的根本缘由在于其对人与自然关系的重新审视，而促成它产生和发展的现实缘由则主要是20世纪百年中的生态危机，表现为全球面临的生态困境、工业社会的精神失落以及由此而蓬勃兴起的绿色运动。其核心观点——"有机论自然观"❸主要观点包括：

---

❶ 见《赫尔佐格与德梅隆全集》.〔德〕格哈德·马克编著，吴志宏译，梁蕾校.中国建筑工业出版社：P210.

❷ 见《国外著名建筑师丛书——凯文·罗奇》徐力，郑虹著.中国建筑工业出版社：P46.

❸ 见《自然之思——西方生态伦理思想探究》.曾建平著.中国社会科学出版社：P133、P141.

图 2　自然建构之美

1）自然是一个生命体。

2）视生命为神圣。

3）把人作为自然的一部分，反对自然只为人类存在。

4）把自然界万物看成是有机关联体，看重自然的整体价值。

"有机论自然观"认为：自然不仅是有机体而且还是一个动态的过程。自然的过程不仅是一个循环或有节奏的变化，而是一个创造性的推进，有机体在经历或寻求一个进化的过程中，每一部分都不断获得并产生新形式。"每一块物质都是自我容涵（self-contained）的，处在某个区域之中和某种空间关系中被动的、固有的网络之中，紧密结合在一个统一的从无限到无限、从永恒到永恒的关系系统之中"。显见，大自然的各个部分就如同一个生物机体内部一样，是如此紧密地相互依赖、如此严密地编织成一张唯一的存在之网，以致没有哪部分能够单独被抽出来而不改变其自身特征和整体特征。这就是自然的"相互依存宣言"❶。

以此为背景，绿色建筑、生态建筑、可持续发展等理念成为共识，并且越来越多地被运用到当代设计中。标志性事件就是2000年德国汉诺威世界博览会委托建筑师威廉·麦克唐纳（William McDonough）发布了著名的《汉诺威原则》（The Hannover Principles），该原则被人们视为生态建筑运动的宣言。建筑领域对"绿色建筑""有机建筑"的重新关注反映了现代化都市在工业化进程中，从创造与生产的根源上，对人与自然关系的又一次深刻反思。

5. 东方传统生态思想给我们的启发：

1）儒家的"天人合一"——仁者以天地万物为一体。

2）道家的"天与人一"——天地与我并生，万物与我为一。

3）佛教的"人境无碍"——天地与我同根，万物与我一体。

东方文化中所表现的天和地、人与物、道德与自然相和谐的哲学观及生态智慧，与西方的生态伦理思想，共同成为我们建筑师学习自然、研究人与建筑、环境、自然关系的智慧源泉。

---

❶ 见《自然之思——西方生态伦理思想探究》．曾建平著．中国社会科学出版：P133、P141．

四、"蔓·设计"在主张什么

"蔓"辞海中的释义是：其五行属木，呈现为立体的蔓延（"曼"为直线的蔓延，"漫"为水平的蔓延），"蔓"本身就洋溢着自然界的勃勃生机。由于"蔓"字非常贴切地浓缩了我们的想法，于是我们就选择了用"蔓·设计"来代表这一理论（图3～图5）。她包含两个维度：

1. 哲学维度——生长：寻找场地已经存在的自然场力，发现场地之中存在的深层结构而不是表象，让建构的建筑呈现出来的状态像自然界一样的有机。

2. 方法论维度——蔓延：把上述设计哲学蔓延到建筑建构的全过程、全类型，并且要调动所有方法和手段打造高完成度的有机建筑作品。

3. 以上两个维度的交集——有机建构，系作为该理论的目标，强调要将所有参与其中的要素非常诗意地组合在一起。

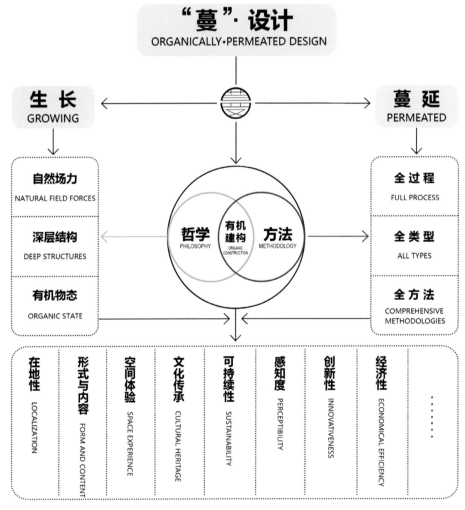

维度1：生长——哲学维度；维度2：蔓延——方法论维度

图3 "蔓·设计"图示

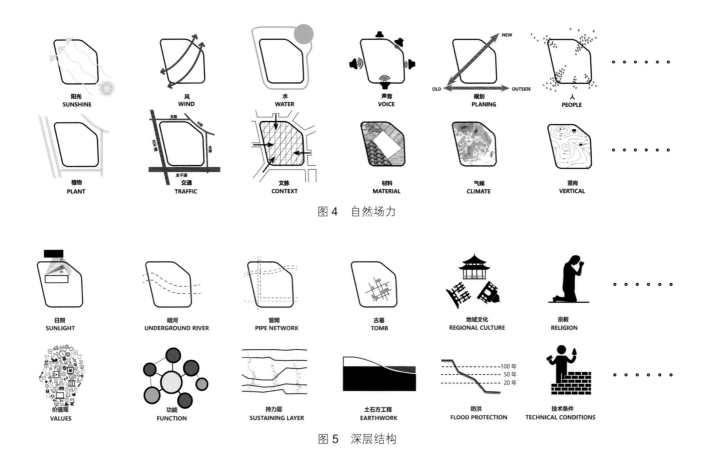

图 4　自然场力

图 5　深层结构

图3～图5解释词：

1）自然场力：包括项目当地的气候、自然资源、交通、上位规划、文脉、使用者的财务状况、材料供应、建造技术水平等。

2）深层结构：这是更为关键的因素，因为它们是隐匿在表象之下，挖掘工作具有更大的难度，因此需要付出更多的精力。例如，甲方的价值观、场地的持力层分布，在遗迹较多的地区，还需要进行地下古墓的挖掘勘探以确定对项目的影响，对于国外客户还需要关注他们的宗教禁忌等因素。

3）有机物态：是自然中所有的组成要素应具备的状态，包括甲方、建筑师和最后完成的建筑都应该是有机物态的一部分。当然这里指的是好的建筑，他们都是有生命的、会新陈代谢的，并且彼此之间互相支持、可持续共生（汉诺威原则第一条）。

4）全过程：包括从策划到后评估的整个闭环全过程。涵盖了项目实施的所有环节，例如施工现场的设计控制。全类型：是指工业和民用以及其他新类型建筑。全方法：是倡导从二维到三维、从平面表达到实体模型和数字模型的设计方法，包括BIM（建筑信息模型）、AI（人工智能）等计算机辅助手段。

在此需要强调的是：

1."蔓·设计"主张建筑在介入场地时要发现其中的自然场力和挖掘深层结构，强调在地性和建筑的生成逻辑。

2."蔓·设计"不是主张把建筑都做得模仿自然，而是要像自然那样富有建构的逻辑和秩序美，强调要学习"自然的内在属性"。

3."蔓·设计"是一个与时俱进的、开放的理论，本身就应该具有生长性。

所以说："蔓·设计"是一个概念并不复杂的理论。

五、介入——一种设计姿态和原则

从"蔓·设计"的两个维度展开来，让建筑以后来者的姿态介入到场地中，是"蔓·设计"所倡导的原则。

一）对大自然而言，我们都是后来者

建筑师的所有行为和成果皆为介入。建筑介入场地应该有"责任心"地去适应现有文脉要求并作出相应的反应。这里所说的"文脉"是指城市里的街道、广场、建筑物、城市历史、文化背景，也可能是场地中或周围自然物态的存在。一旦建筑介入自然环境，就应该与场地中及周围已有的有机物态建立关系，这种关系的质量取决于建筑师，需要建筑师借助自己的知识、灵感和智慧去进行选择和取舍，以确保与场地既有物态保持一个融入关系，避免生硬或疏远。

二）关注介入的后果

1962年美国作家雷切尔·卡森（Rachel Carson）在其著作《寂静的春天》中指出人与自然的关系除了人要适应自然外，还有另一面，那就是人可以极大地影响自然环境，而这一点在此之前被很多人所忽视。

建筑师时刻要意识到由于新建筑的介入，对环境的影响程度经常会超出我们的预想，其会给城市环境带来一些不可忽视的严重问题。我们应该知道，新建筑的介入将会决定人与建筑、建筑与城市相互作用的方式，不管是好是坏，这必然是一种相互作用而不是简单的单方面的反应。建筑师的行为总归慎重一点为好，不要轻率地决定城市或者景观的命运和发展前途。

三）介入的途径和方法

毫无疑问，这种介入应是在充分发现和挖掘场地的自然场力与深层结构基础上，采用相应被动与主动的方式介入其中。当然这里鼓励建筑师关注积极意义上的介入，即如何认识已有自然场力和深层结构的价值，如何在新建筑介入过程中有责任心地去适应已有因素的要求并作出反应。因此，新建筑介入的方式应该是极其丰富多彩的。罗伯特·文丘里（Robert Venturi）在其伦敦国家画廊新馆、贝聿铭在其华盛顿国家美术馆东馆、马里奥·博塔（Mario Botta）在瑞士的圣维塔尔住宅，都因为谨慎同时主动的介入造就了成功的案例。

四）建筑师应敏锐地发现并挖掘与建筑有关的构成要素，才有可能建造出"敏感的容器"[1]

美和秩序存在于自然天成的事物中，它们具有隐匿性、多样性和多变性，需要调动建筑师的意识和感官（如触觉、嗅觉、味觉等）去感受和捕捉场地、空间、材料、纹理、色彩等因素，我们戏称为"用考古学的角度"来观察项目的场地和环境，需要对建筑的魅力、气氛、活力和风采所带来的影响做出评估。好设计的感染力来源于我们本身以及我们对感性世界和理性世界的认知能力，建筑师只有精心挖掘才有可能创作出如彼得·卒姆托（Peter Zumthor）所说"敏感的容器"。当这个容器在回应外界时如果足够敏感的话，她就会自然呈现出一种美的品质。因此，任何项目都应该是没有预先固定的建筑样式，而是尽力从设计任务书、场地、材料等要素中生成的基础问题，在分析解决这些问题过程中完成了建筑的创作。正因为如此，对于好的建筑，我们很难给他们的品质找到一个统一的特征，如果我们非要加以概括，那就会剥夺每一个建筑各自的特点和品质。

---

[1] 见《思考建筑》. 彼得·卒姆托著，张宇译. 中国建筑工业出版社：P12.

六、"生长与蔓延"两个维度的交集——有机建构

"建构"是指要诗意化的表达建筑的联结关系，此处所说的"有机建构"就是要向自然学习，追求自然天成的建构状态，还有一层递进的意思是说，建筑在建造过程中要寻求与场地的建构，让建筑像从场地中生长出来、牢牢扎根于此。

肯尼思·弗兰姆普敦（Kenneth Frampton）等提出的建构理论，一是要抵抗技术理性的绝对控制，二是抵抗商品社会所导致的"建筑图像化的侵蚀"[1]。19世纪德国建筑理论家卡尔·博迪舍（Karl Botticher）认为建筑的技术性要素与艺术性表现之间应建立起直接的桥梁——"建构"：即给予艺术品质及文化内涵以坚实的技术基础[2]。在此需要强调的是：

1. 建筑不同于雕塑，雕塑由于尺度的原因强调无缝连接，而建筑由于尺度庞大而取决于连接的质量。

2. 有机建构：首先是一个过程，然后才是一个结果，所以就某一个表现形式或者结果而言，是本末倒置的。其中拼接的逻辑、隐匿在其中的几何布局、材质之间的冲突、承重与支撑的区分、纹理的组合等，都需要建筑师付出精力去反复研究和推敲。

3. 只有当我们一步步成功解决了从场地、材料、意图中产生的问题之后，令人惊讶的结构和空间才能呈现，我相信这样的结构和空间才会拥有一种潜在的原生力量。因为建筑应该准确真诚地反映出自然场力和深层结构的作用，她应是两者作用后的结果。

七、好建筑的标准

什么是好建筑？这是经常被问也经常自问的问题。"蔓·设计"理论解决了篇首说的"怎么看待建筑"和"如何表达建筑"问题之后，好建筑的标准也就自然产生。

一）良好的在地性

从强调建筑与当地自然和人文环境相契合，到强调项目所在地区的地域独特性。发现自然场力、挖掘深层结构，以"介入"的原则展开项目，将人、建筑与环境紧密地联系在一起，与当地建造智慧、风土人情、生活习俗都有关联。艺术想象加理性思维的创作过程应是从外至内的和从内至外的分析过程，针对不同的环境建筑师应用不同的作品来解答。

二）形式和内容统一

1. 人们往往容易迷恋形式并予以夸大，总是大肆渲染一些与建筑本质无关紧要的东西，建筑师可以掀起一次大讨论，以抵制那些无用的形式。

2. 屈米说：建筑不是一种形式的学科，而是一种学科的形式。

3. 如果只讲形式，那就成了单纯的形式智力游戏。建筑不是一个简单的视觉形象，它应该是一个空间的场所和记忆。无论她显赫抑或平凡，一定会有人在其中留下生活的痕迹和回忆。

2010年上海世博会沙特馆（以下简称沙特馆）如没有海上丝路的寓意也就不会是船的构想，自然也不会在内部生成壳型全景融入式的流线设计。内容和形式谁决定谁已不重要，重要的是内容与形式的统一，并能高度融合，产生出一种在两者之上的新的设计语言。

三）空间体验丰富

空间是建筑的主角。老子在两千五百多年前讲到"有无相生"理论的时候，已经认识到做建筑其实是在做空间。我在做沙特馆的时候，觉得我们的先贤太伟大了，空间是主角，而在此应该强调的是人又是空间的主角，一个空间再宏大，实际上是靠人来体验的。这就是挪威哲学家克里斯蒂安·诺伯格-舒尔茨（Christian Norberg-Schulz）所提出的"场所精神"，场所强调的是要人来体验，人和空间组合在一起才能产生真正的建筑体验。虽然现在有人说，当代建筑已进入一个自由时代或者说试验的时代，但不管怎样，空间是建筑的主角，人又是空间的主角，这样一个命题应该是永远不变的。

---

[1] 见《建构文化研究》.〔美〕肯尼思.弗兰姆普敦著,王骏阳译.中国建筑工业出版社.

[2] 见《当代建筑理论》.青锋著.中国建筑工业出版社：P268.

四）好的感知度

"感知度"是信息论中的一个名词，是指人对信息的接近程度。2010年上海世博会上，为什么观众会追捧沙特馆，说明它的感知度高。从《一千零一夜》故事和海上丝绸之路历史提取出的元素，让大家都非常熟悉，包括国外同行他们小时候也都看过《一千零一夜》。一个建筑的构思阶段，寻找一个高感知度的元素和理念，更能引起共鸣，建筑的概念要避免晦涩。

建筑师不应把建筑当成塑造其个人特征的手段。由于其强烈的社会属性，故不能用评价其他艺术形式如雕塑、绘画的方法来评价建筑，它还是一种空间的艺术，负有很多文化的使命。建筑已经脱离了纯个人的美学范畴，因为对建筑的价值判断已经融入了业主、包括大众等各个层面人士的意见。

五）关照文化传承

建筑师应该深知自己负有的使命，特别是在一些具有深厚历史积淀的城市、地区设计建筑，如何积极地关照、延续和传承地域文化，是必须面对的一个课题。

六）可持续性

1. 可持续性不是一个口号，它主要强调思想和技术，它是一种全面的建筑观和超越标准的建筑的DNA，不是一种建筑流派思想，不是纯粹强调形象和风格。

2. 它既要考虑当下，又要兼顾全生命周期，而不仅仅指设计阶段。

3. 始终要思考怎样借助自然的恩惠，对一切自然资源加以充分利用。它是指利用对周围生态系统或社区环境无害的设计方法、材料、能源以及可开发空间，并遵守社会、经济和生态的可持续原则，目标是减少建筑能耗以及对环境的负面影响。它分为"低碳、零碳、负碳"三种绿色建筑标准。具体的方式可以根据项目的实际条件进行选择。如被动与主动、可再生能源系统、绿色材料和海绵城市等（图6）。

七）经济性

经济性强调建筑的成本效益，这是建筑得以实现的现实依据。经济性不只限于简单地节约造价，还要从首期投入与长远运行的综合性评价，强调要从建筑的全生命周期看待建筑经济的合理性。

八）创新性

创新是建筑师职业生涯必须面对的无尽的前沿[1]、永恒的课题。这句话说得非常深刻：设计是对社会变革的一种回应[2]。沙特馆的参观方式观众之所以比较喜欢，正是因为这种展示的方式呼应了高科技时代观众参观博物馆时候的一种需求。它颠覆了传统二元并置式的观看模式，创造了一种更逼真、更融入或者说更震撼的参观体验。从这点上来说，创作不应把太多的精力花在外形的时尚上，而是应在空间体验上，时尚的东西是容易过时的，本质的东西才是永恒的。

当然，好建筑的标准应该不止这些，随着我们研究的深入，相信还会加入一些共识进来。

八、"蔓·设计"下的"大建筑观"

一）客观背景

1. 进入新时代，中国进入由从求量转向求质的发展阶段。

2. 房地产减速的同时，高科技工业项目、IT项目剧增。

3. 信息时代（智能时代）的新建筑类型越发体现出介于民用建筑和工业建筑之间的"中间特性"，民用建筑与工业建筑之间的界限更加模糊。

二）主观背景

1. 从彼特·贝伦斯（Peter Behrens）领导的现代主义建筑的发展过程，反思工业与民用二元分类法带来的局限性。

2. 从本人所在的中国电子院的"基因"、业务内容以及

---

[1] 见《合和建筑观》.陈雄著.中国建筑工业出版社：P14.
[2] 见《认知：设计意味着商机》.〔美〕保尔.海施编著，杨慧鸣译.京华出版社：P38.

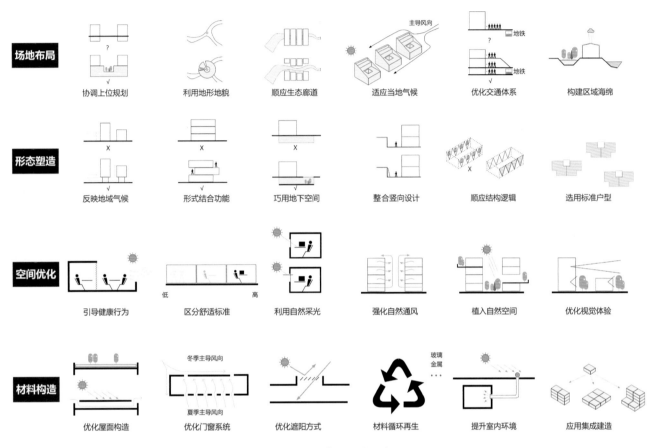

图 6 绿色建筑设计方法

建筑师的人员总量和构成,来反思将工业建筑与民用建筑做二元划分所带来的局限性。

三)"大建筑观"的定义及其与"广义建筑学"的区别

1."广义建筑学"定义:就是通过城市设计的核心作用,从观念和理论基础上把建筑学、地景学、城市规划学的要点整合为一,对建筑的本真进行综合性地追寻。

2."大建筑观"定义:随着信息时代的到来,传统的分类方法越发显得简单和粗糙,本概念就是打破和取消传统的、简单的工业建筑与民用建筑之分,将只要满足人类活动需要的物理和精神空间皆看为建筑,是为"大建筑观"。

四)"大建筑观"提出对建筑师的意义

在阐述"大建筑观"后,以中国电子院为例,我们有必要进一步来探讨为什么不能再去简单、刻意地划分建筑工业与民用建筑:

1. 溯源西方近 150 年的近现代建筑发展史,从一开始就没有将工业和民用概念分开。现代建筑先驱彼特·贝伦斯(Peter Behrens)第一个现代建筑开山之作就是工业建筑——柏林通用电气(AEG)透平机车间,他的门徒现代建筑大师格罗皮乌斯(Walter Gropius)、密斯(Mies Van der Rohe)、柯布西耶(Le Corbusier)也都是以工业建筑作品开启了自己的职业生涯,纵横于工业建筑和民用建筑领域。

2. 进入当下信息时代或者说智能化社会,新的建筑类型不断涌现,无疑是在印证这种绝对的、生硬的工业建筑与民用建筑分类已不合适。

3. 从学术上讲,本质上所有建筑类型都有工艺设计存在。例如,芯片厂、医院、酒店、博物馆、数据中心都有工艺，仅仅是功能类型不同而已，对空间体验、形式美及其与功能的平衡要求程度不同而已。从建筑学的本质来讲都是一样的。

4. 目前，建筑设计市场处在"僧多粥少"的状态，设计院处于被选择的地位，所以市场决定了：设计院应该也必须保持一个开放的心态，有什么就做什么。

5. 我国已经从"有没有"进入"好不好"的发展阶段，业主对工业建筑特别是高科技工业建筑的总体要求（包括空间环境质量、艺术性等）已经不亚于民用建筑。

6. 况且，从客户自身发展看，原来很多从事工业生产的客户现在也开始转型做非工业生产的项目，如果要把这种信任、服务延续下去的话，我们必须是多面手。

7. 现在行业倡导"全过程咨询和设计""建筑师负责制"，更是要求建筑师能够全过程、全专业负责把控，这就需要工业院的建筑师打破既有工作界限，具有把握全局、全过程的大建筑观。

8. 从中国电子院的历史和业绩积累、目前建筑师人员构成情况以及由此被决定的在中国设计行业的民用建筑的"江湖位次"来看，在这种条件下，仍要刻意强调工业建筑和民用建筑之分，无疑是作茧自缚，特别是本来就不强大的建筑师队伍还要截然分开，变得都不强大，到头来两种类型都做不好。所以，只有攥成一个拳头打出去才会有力量。

综上所述，遵循建筑学基本要素及其外延要求，是所有建筑类型创作中建筑师必须遵循的基本职业准则，而类型只是具体设计内容不同而已。摒弃惯性的、简单化的二元分类做法，消除对建筑类型的成见，中国电子院定会抓住当前的市场契机，迎来更新更大的发展。所谓"大建筑观"，是超内容、超类型的大建筑，大建筑是一种格局和气势，是建筑师的选择和价值取向。

"蔓·设计"是关于于我们的设计哲学和方法论的理论构架，是我们实现"有机建筑"建构的主导原则。基于工业院特点的"大建筑观"和基于对建筑学深刻理解的"蔓·设计"两者的结合，未来一定会为我们建筑师、工程师永续深耕原有优势领域并开拓新领域，奠定扎实的发展基础。

结语

信息时代使建筑师获得资讯变得越来越便捷，AI的发展使我们的表达和展示手段也越来越多样。但现实是，当下中国建筑的目标并没有越来越清晰和坚定，而这种状态正好证明了建立我们自己理念的重要性。

"蔓·设计"理论倡导要用整合的观点去分析、判断事物，要求要尽可能准确地把握事物的内部结构和本质。正是因为她强调用辩证、理性的思维方法，去挖掘每个项目的自然场力和深层结构，所以作品的表现形式应该是多元变化的。建筑创作虽然表现为诸多矛盾的交锋、融合、重组，但这种交锋、融合、重组不仅仅是停留在审美和智力游戏层面，其中的文化、社会含量更应占其首位，我们对建筑进行探索，不仅仅是为了寻找新意和"手法主义"审美的可能性，更是要用这种方法寻找建筑的本质意义。

"蔓·设计"理论在回答了"怎样看待建筑"以及"如何表达建筑"这两个核心问题后，未来还有大量的工作要做。对经典设计手法和优秀建构细节的研究、学习，以及在此基础上的创新，还需要我们投入更大的精力。这里需要特别指出的是，即使在理念相同的情况下，若建筑表达方法不同，也会呈现出完全不同的结果，反之亦然。当然，如何看待建筑这件事决定了建筑师的高度，而手法和技巧显然属于第二层次。

建筑师往往更喜欢谈论他们作品中理性和慎思的一面，而较少谈起曾经启发设计的激情闪现，毫无疑问，在理念确立后，方法论上还有大量工作要做。那接下来建筑师的"手法可能性的探索""新切入点的寻找"，特别是"艺术想象力的激发"等方面，就成了我们创新未来无限潜能的关键。

作为承上启下的一代，我们必须保持清醒、保持实践与思考并重的状态，创作中国当代建筑的使命还在肩上，没有紧迫感和危机感就无法成长，同时也意味着将面临更大的危机。

2024.5.30 于北京五路居

The Propositions of "Organically-Permeated Design"

Introduction

"Organically-Permeated Design" represents a design philosophy that I articulated in 2017, derived from over three decades of reflection, practice, and refinement in architectural design, further tested through a series of practical experiments in my studio. Guided by this unified vision, our studio has embarked on a prolific period of creative activities, producing numerous works. Simultaneously, we have actively participated in exhibitions and academic exchanges externally, while internally strengthening our learning processes. This involves retrospection and discussions concerning practical issues encountered in projects. We continuously supplement, revise, and improve this concept, and this work will continue.

Over the course of seven years, we have collectively experienced the unforgettable "public health emergency". As a result, many projects commissioned to our studio have been delayed or canceled. However, this is of secondary concern. It is most imperative that: reality increasingly compels architects to earnestly reexamine the relationships between humans and nature, architecture and nature, as well as what constitutes organic architecture."Organic architecture" is certainly not a new concept; predecessors and masters such as Wright, Le Corbusier, Gaudí, Howard, Alvaro Siza, KenzoTange, Pei, Calatrava, Toyo Ito, RCR, and D. Pearson, have tirelessly explored it in both practice and theory. This is because of the inherent complexity, diversity of forms, and vast systems of nature, along with its particular significance to architects. It serves as the fundamental site for architectural creation; how to view and handle the relationship with nature is the primary task an architect must continuously confront.

This article aims to present a periodic report, during the process of exploring the aforementioned issues, on our re-evaluation of the fundamental elements of architecture, as well as our philosophical and methodological considerations and practices regarding "how to perceive and express architecture", making a periodic report to everyone.

1 Origin—Anxiety in Practice and Theory

1.1 Origin I—The Need for Practice
1.1.1 Clarifying my own thoughts
2016 marks the 30th anniversary since my undergraduate graduation. It seems that for quite some time, the theoretical anxiety about architecture has been troubling me alongside the progress of my practice. Therefore, it is urgent at this juncture to clarify and summarize the cognitive system of the profession I am engaged in, to make a complete and systematic clarification of the ontology of architecture. Faced with the diversity and complexity of contemporary architectural theory, as well as the various conflicting ideas within the industry, we must discover the truth of architecture and hope to somehow verify it in practice.
1.1.2 Making myself clear to the youngsters
After the establishment of the studio, especially between 2013 and 2015, with the expansion of our business, the studio grew from 6 people to nearly 20 people. Faced with the influx of young architects, it was necessary for me to succinctly and clearly tell them how exactly to practice architecture; what is good architecture; how to grasp the key points among the complex elements. The team urgently needs to reach a consensus. In the process, I also try to fulfill the desire to break the old notion of "architecture can only be learned, not taught" that I had when I was at school.
1.1.3 Why does a team need an explicit basic consensus?
(1) Adhering to cognition: The information explosion in the Internet era leads to the more you know, the more you lose, ultimately resulting in self-loss. Therefore, team

members need to have a clear understanding and maintain identification ability.
(2) Improving efficiency: Architectural creation is teamwork. Establishing a consensus minimizes unnecessary detours and indecision, which in turn, enhances efficiency.
(3) Dealing with external pressures and interferences from clients, management departments, etc.: As clients' insights become broader and they become more assertive and confident, architects (especially young architects), without firm professional principles, are prone to becoming various styles of "money printing machines" for clients.

1.2 Origin II—Scholarly Reflections on Architecture
1.2.1 The relationship between humans, architecture, sites, and the broader environment, extending to nature itself, constitutes a perpetual area of contemplation and challenge for architects.
Humans have always fancied themselves as the most perfect structure in the world, and have demanded perfection from everything around them. A trait that has only intensified as human capabilities have expanded uncontrollably. Architecture, as a byproduct of human activity, vividly reflects these behavioral patterns. Cities and buildings, as artificial constructs, have evolved through this relentless pursuit of perfection. Regrettably, while showcasing human strength and greatness, the outcomes presented by architecture and urban development are not entirely satisfactory. An increasing discord and confrontation between man-made structures and the natural environment are evident, and these conflicts are intensifying. Nature gave birth to humans, who in turn have created buildings, modern cities, and modern lifestyles. Unfortunately, modern life alienates us increasingly from nature and has led to numerous misinterpretations of it, consequently depriving us of opportunities to truly grasp the essence of this energy.
1.2.2 In the Western narrative, human history has often been depicted as one of a struggle against and conquest over nature. However, this perception has proven problematic, as such antithesis has brought significant challenges, causing us to overlook our vital connection with the environment and to erroneously believe that we can transcend natural laws and be immune from punishment by the natural world, a fate from which other species cannot escape. The emergence and conclusion of the "public health emergercy" compel us to engage in deep reflection once again.
1.2.3 The plurality and complexity of contemporary architectural theories — ranging from traditional Modernist doctrines to new theoretical adaptations — present a landscape of diverse and often conflicting viewpoints. With an overflow of information, architects are put to the test, and they urgently need to have their own philosophy to uphold and guide their practice, like taking a reassuring tonic, so as to avoid conceptual entanglements. Otherwise, it is likely to result in simply adopting others' ideas as new and repeating them like a parrot.
1.2.4 Empirical evidence suggests that architects lacking a solid conceptual foundation can easily fall into messy cognitive mires, leading to disorganized and chaotic methodologies in the mind, ultimately preventing him or her from achieving remarkable attainment in the field. Japanese architecture, which balances heritage with innovative thinking and is celebrated for its distinctive design philosophies and styles, exemplifies this, as evidenced by Japanese architects making up one-sixth of the Pritzker Prize laureates.

2 Review of the Constitutive Elements of Architecture

The elements of architecture encompass a wide range: To clarify the core issues of architecture is essentially to categorize and summarize these elements to distinguish primary from secondary concerns and to establish priorities. By examining the essence through observable phenomena, we have categorized the components of architecture into three distinct levels:

2.1 Autonomy in Architecture
It includes site, space, form, structure, materials, proportions, colors, etc. It can be internally legislated, independent of other elements outside the domain. It is also the intrinsic element upon which the establishment of architecture depends; nothing would be possible without internality.

2.2 Extensiveness of Architecture
This includes:
2.2.1 New materials and technologies; or it can focus on the changes in construction and usage methods caused by new technologies, materials, and social environmental issues.
2.2.2 Social elements—politics, economics, culture, and population;
2.2.3 Humanities —philosophy, environmental studies, sociology, anthropology, etc. —utilize the development of

other disciplines to research architectural theory.

2.3 Expression in Architecture: This involvies style, technique, architectural language, etc.

Regarding the expression in architecture, such as the choice of techniques, various attitudes, and methods can serve the construction of architecture concerning the same elements within natural forces, including preservation, reflection, filtration, isolation, dilution, and others.

To elucidate this matter, one can consider the example of Zaha Hadid and David Chipperfield. Both were classmates at the Architectural Association School in London, and both achieved remarkable success. They also attempted to establish relationships with the urban context and neighboring structures in their designs, seemingly sharing similar philosophies. However, due to their divergent approaches to architectural expression, they manifest vastly different styles.

In my opinion, in the contemporary Chinese context, discussing and delineating the constitutive elements and levels of architecture not only serves the factors mentioned earlier but also holds particular and pressing practical significance. This is due to several factors: on the one hand, the traditional and fundamental core of modern architectural studies has not yet been fully established in China, and the extensional factors of architecture exert a more pronounced influence on architectural creation in China. On the other hand, with China's reform and opening up, international architects have gained unprecedented opportunities to realize their ceative work in China. Contemporary Chinese architecture has thus become intertwined with the global architectural dialogue. In this context, a critical challenge for Chinese architects is how to remain grounded and clear-headed, returning to the essence of architecture itself. This involves re-evaluating these themes and elements from a fresh perspective and considering their place and transformation within contemporary architectural discourse. This task is essential for the development of Chinese architecture today.

3 Why the Emphasis on Learning from Nature

3.1 Whenever we conduct site surveys, whether we carefully overlook the site or scrutinize every brick, stone, grass, and tree, it is not difficult to find that they are filled with the harmonious unity of natural forces. They all exhibit their order, which is meticulously organized and planned, and each detail displays the beauty of construction. The beauty of natural order and the beauty of harmony in the unity of form and content are captivating and moving. This phenomenon is referred to as "The Hidden Geometry of Nature" by J. Herzog. He states: "With this, we mean the complexity of a system of relationships which exists in nature, possessing an unatainable perfection that cannot be clearly explored, and whose analogy in the realm of art and society interests us. Our interest is thus the hidden geometry of nature, a spiritual principle and not primarily the outer appearance of nature."❶

The German poet Novalis once said, "The secret of art is to turn every natural phenomenon and natural order into a rule—similar to the art of construction." In the 1970s, when asked how he would design the IBM Museum, Kevin Roche replied: "We want to establish an internal display of high technology and its complexity, but this display is somehow connected to the natural environment, as nature possesses the ultimate complexity."❷

3.2 We reemphasize learning from nature, I believe, for the following reasons:

3.2.1 The laws of balance in nature and the functioning of ecosystems are the basis for human understanding and application of scientific principles. Perception and experience of nature are the foundation of human cognitive system.

3.2.2 Nature has always been the source of human inspiration. The diversity and complexity of the natural world indeed inspire human—including architect's—creativity, especially in artistic creation.

3.2.3 The stark contrast between the ugliness of many human-made buildings and the enchanting beauty of natural order proves that human achievements and capabilities have not surpassed nature. Therefore, we should learn from nature.

3.2.4 The formation of Western ecological ethics provides us with inspiration:

The fundamental reason for the formation of this theory lies in its re-examination of the relationship between man and nature, driven primarily by the ecological crises of the 20th century, manifested in global ecological challenges, the spiritual disillusionment of industrial societies, and the flourishing green movement. Its core viewpoint—the "or-

---

❶ "Herzog & De Meuron-The Complete Works" edited by Gerhard Mack, translated by Wu Zhihong, and proofread by Liang Lei, published by China Architecture & Building Press, p.210.

❷ "Famous Foreign Architects Series - Kevin Roche" by Xu Li & Zheng Hong, published by China Architecture & Building Press, p.46.

ganic view of nature"❶—includes:
1) Nature is a living organism.
2) Treating life as sacred.
3) Considering man as part of nature, opposing the notion that nature exists solely for man.
4) Viewing all elements of nature as an organic alliance and valuing the overall value of nature.

"Organic Naturism" posits that nature is not only organic but also a dynamic process. The natural process entails more than just cyclical or rhythmic changes; it represents a creative progression wherein each part of the organism continually acquires and generates new forms while undergoing or seeking an evolutionary process. "Every piece of matter is self-contained, passively nestled within a particular region and spatial relationship, tightly integrated into a unified system of relations from infinity to infinity, from eternity to eternity." It is evident that various parts of nature, much like within a biological organism, are so closely interdependent and intricately woven into a single web of existence that no part can be singled out without altering both its individual and overall characteristics. This is the "declaration of interdependence" in nature❷.

Against this backdrop, concepts such as green architecture, ecological architecture, and sustainable development have become consensus and are increasingly being applied in contemporary designs. A landmark event was the 2000 Hanover Expo in Germany, where architect W. McDonough released the famous "The Hannover Principles," seen as a manifesto of the ecological architecture movement. The renewed focus in the architectural field on "green building" and "organic architecture" reflects a profound reevaluation of the relationship between man and nature from the roots of creation and production in the modernized urban context amidst industrialization.

3.2.5 Inspiration from Eastern traditional ecological philosophies:
1) Confucianism's "Heaven and man are united as one" — the benevolent and virtuous treat heaven, earth, and all things as a unity.
2) Taoism's "Heaven and I are of the same root" — heaven and earth coexist with me, and all things are combined with me.
3) Buddhism's "Unity of Self and Environment" — heaven and earth share the same origin with me, and all things are interconnected with me.

---

❶❷ "Natural Thinking: Exploring Western Ecological Ethics" by Zeng Jianping, published by China Social Sciences Press, p.133, p.141.

The philosophical views and ecological wisdom expressed in Eastern culture regarding the harmony between heaven and earth, humans and objects, morality and nature, along with Western ecological ethical thoughts, collectively serve as a wellspring of wisdom for architects to study nature and research the relationship between humans, architecture, environment, and nature.

4 Propositions of "Organically-Permeated Design"

4.1 The term "*Organically-Permeated*" in the CIHAI dictionary refers to its botanical association with wood, presenting itself as a three-dimensional extension (" 曼 Stretch" for linear extension, " 漫 Overflow" for horizontal extension). "*Organically-Permeated*" itself embodies the vitality of the natural world. As this word aptly encapsulates our ideas, we chose to use "*Organically-Permeated Design*" to represent this theory. It encompasses two dimensions:

4.1.1 Philosophical Dimension—Growing: Seeking the existing natural forces within the site, discovering the underlying structures rather than surface appearances, and allowing the constructed architecture to exhibit an organic state akin to the natural world.

4.1.2 Methodological Dimension—Permeated: Extending the aforementioned design philosophy to the entire process and typology of architectural construction, mobilizing all methods and means to create highly organic architectural works with a high degree of completion.

4.1.3 The intersection of the above two dimensions—organic construction - is the goal of this theory, emphasizing the poetic combination of all the elements involved.

4.2 It is important to emphasize the following:

4.2.1 *"Organically-Permeated Design"* advocates that architecture, when intervening in a site, should discover natural forces and excavate deep structures within it, emphasizing the localization and the logic of architectural generation.

4.2.2 *"Organically-Permeated Design"* does not advocate for merely mimicking nature in architecture but rather seeks to imbue architecture with the constructive logic and ordered beauty found in nature, emphasizing the need to study the "inherent attributes of nature".

4.2.3 *"Organically-Permeated Design"* - Architecture is a progressive and open theory and should inherently possess a sense of growth.

Therefore, *"Organically-Permeated Design"* is a concept that is not overly complex.

5 Intervention in Design: A Stance and Principle

Emerging from the two dimensions of "*Organically-Permeated Design*", the principle advocates for the creation of projects in which architecture intervenes as a later arrival. This principle is expounded in several key points:

5.1 To nature, we are all latecomers

All actions and outcomes of architects are "interventions". New buildings intervening in a site should responsibly adapt to and respond to the existing cultural and contextual demands. The "context" mentioned here could refer to urban streets, squares, buildings, city history, cultural backgrounds, or the presence of natural elements in or around the site. Once a new building intervenes in the natural environment, it should establish a relationship with the pre-existing organic state of nature in and around the site. The quality of this relationship is contingent upon the architect's ability to leverage knowledge, inspiration, and wisdom to make choices that ensure an integrated relationship with the site's inherent qualities rather than creating a relationship that is forced or alienated.

5.2 Caring for the consequences of intervention

In her seminal work "Silent Spring" (1962), American author Rachel Louise Carson highlighted the relationship between humans and nature, noting not only the necessity for humans to adapt to nature but also their significant impact on the natural environment, an aspect previously underestimated by many.

Architects should always be aware that the degree of "addition" or impact due to new architectural interventions often exceeds anticipated levels, potentially bringing severe issues to the city and environment that cannot be ignored. Architects need to recognize that their interventions will determine the modes of interaction between people and buildings, and between buildings and cities, for better or worse. This interaction is inevitably mutual rather than a mere unilateral response, thus necessitating cautious and responsible architectural behavior that doesn't recklessly dictate the fate and developmental trajectory of cities or landscapes.

5.3 Methods of Intervention

Undoubtedly, such interventions should be based on a thorough discovery and exploration of the site's natural forces and subsurface structures, employing both passive and active approaches. Emphasis is placed on positive interventions—how to recognize the value of existing natural field forces and deep structures, and how, in the process of new architectural interventions, to responsibly adapt to and respond to these existing factors. Consequently, the methods of new architectural interventions should be richly varied and contextually sensitive. Success stories of cautious yet proactive interventions include the new wing of the National Gallery in London by Venturi, the East Building of the National Gallery of Art in Washington by I.M. Pei, and Botta's House Riva Sun Vitale in Switzerland.

5.4 Sensitive Container

Architects should keenly discover and explore the elements related to architecture to create "sensitive containers"[1]

Beauty and order, inherent in natural objects and characterized by their covertness, diversity, and variability, require the mobilization of an architect's awareness and senses (such as touch, smell, and taste) to capture and appreciate the elements such as site, space, materials, textures, colors, etc. We jokingly refer to observing the site and environment of a project with an "archaeological perspective", which requires evaluating the impact of the building's charm, atmosphere, vitality, and charm. The infective power of good design lies in our own understanding of the sensual and rational worlds. Only through meticulous exploration can architects create "sensitive containers", as P. Zumthor put it, which, when responsive enough to external influences, naturally present a beautiful quality. Thus, for any project the architect should not preconceive an architectural style but should strive to originate from fundamental issues derived from the design brief, site, materials, etc., completing the architectural creation through the analysis and resolution of these issues. Consequently, it is challenging to assign a unified characteristic to good architecture; to generalize would strip each building of its individual features and qualities.

6 Organic Tectonics: The Intersection of Growing and Permeated

The concept of "tectonics" refers to a poetic expression of the connections within architecture. The term "organic tectonics" implies learning from nature and striving for a naturally inherent tectonic state. Further, it suggests that

---

[1] "Thinking Architecture" by Peter Zumthor, translated by Zhang Yu, published by China Architecture & Building Press, p. 12.

during the construction process, architecture should seek integration with the site, growing from it and being deeply rooted therein as if it were a natural extension.

Theories of tectonics such as those proposed by Kenneth Frampton oppose the absolute control of technical rationality and resist the "erosion of architectural imagery" ❶ as a result of the commodity society. In the 19th century, German architectural theorist Karl Bötticher believed in establishing a direct bridge between the technical elements of architecture and its artistic expressions—"construction", that is, to provide a solid technical basis for artistic quality and cultural content❷ It should be emphasized here that:

6.1 Unlike sculpture, which emphasizes seamless connections due to its scale, architecture depends on the quality of connections because of its larger scale.

6.2 Organic Tectonics is first a process and then a result. Discussing it merely in terms of a specific form or outcome is to put the cart before the horse. The logic of assembly, the hidden geometrical layout, conflicts between materials, distinctions between load-bearing and bracing, and combinations of textures all involve a process that architects must painstakingly study and deliberate.

6.3 Only when we have successfully resolved issues arising from the site, materials, and intentions can we reveal surprising structure and space. I believe such structure and space will possess a latent and crude power. Because architecture should accurately and sincerely reflect the effects of natural field forces and deep structures, it should be the result of their interaction.

7 Criteria for Good Architecture

What constitutes good architecture? This is a frequently asked and self-posed question. After addressing the initial questions of "how to perceive architecture" and "how to express architecture" presented by the "Organically-Permeated Design" theory, the criteria for good architecture naturally emerge.

7.1 Strong localization:
This involves emphasizing the compatibility of the building with the local natural and cultural environment, as well as emphasizing the regional uniqueness of the project location. By discovering natural forces and exploring deep structures, and by unfolding projects with the principle of "intervention," a close connection is forged between people, architecture, and the environment, which also relates to local building wisdom, customs, and lifestyles.

The creative process, which integrates artistic imagination with rational thought, should entail both an analysis from the outside in and from the inside out. In response to diverse environments, architects should employ distinct works as solutions.

7.2 Unity of Form and Content

1) People are often obsessed with form and exaggerate it, always exaggerating things that are irrelevant to the essence of architecture. Architects can initiate a big discussion to resist useless forms.

2) As Bernard Tschumi once articulated, "Architecture is not a discipline of form, but rather a form of discipline."

3) Focusing solely on form transforms architecture into a mere intellectual game of shapes. Architecture transcends simple visual representation; it ought to embody a space of place and memory. Whether prominent or modest, every building invariably retains traces of life and memories left by its occupants.

If the Saudi Arabian Pavilion does not have the symbolism of the Maritime Silk Road, it would not have the concept of a ship, and naturally, it should not create a shell-shaped panoramic integrated streamlined design inside. It no longer matters who determines what; what is important is the unity of content and form, which should be highly integrated to produce a new design language that transcends both.

7.3 Rich Spatial Experience
Space is the protagonist of architecture. More than 2500 years ago, Laozi discussed the philosophy of "Being and non-being produce each other," recognizing that architecture fundamentally deals with space. While working on the Saudi Pavilion, I deeply admired our predecessors, realizing the profound significance of space. It is essential to emphasize that while space takes center stage, it is ultimately people who play the leading role within it. No matter how grand a space may be, its true essence is experienced through human interaction. This embodies the concept of "genius loci" proposed by Norwegian philosopher N. Schultz, suggesting that places require human presence to truly come alive; it is the combination of people and space that generates genuine architectural

---

❶ "Studies in Tectonic Culture" by Kenneth Frampton, translated by Wang Junyang, published by China Architecture & Building Press.

❷ "Contemporary Architectural Theory" by Qing Feng, published by China Architecture & Building Press, p. 268.

experiences. Although some argue that contemporary architecture has entered an era of freedom, or perhaps an experimental phase, regardless, the proposition that space is the protagonist of architecture, and humans are the protagonist of space, should remain an immutable principle.

7.4 Good Perceptibility
Perceptibility is a term from informatics that refers to an individual's proximity to information. The popularity of the Saudi Pavilion at the 2010 Shanghai World Expo suggests its high perceptibility. Elements derived from the narratives of "Arabian Nights," and the Maritime Silk Road, are widely recognized. International peers have also read "Arabian Nights" in their childhood. During the conceptual phase of architectural design, selecting an element or concept with highly perceived intensity can elicit stronger resonance, while avoiding overly obscure elements is essential.

Architects should refrain from regarding architecture solely as a tool for cultivating their personal identity. Given its pronounced social attribute, architecture should not be subject to evaluation through the same criteria applied to other artistic forms like sculpture or painting. Rather, it constitutes a spatial art form replete with manifold cultural imperatives. Architecture has transcended the confines of purely individual aesthetics, as assessments of its value now incorporate the perspectives and judgments of stakeholders, including clients and the broader public across diverse strata.

7.5 Caring for Cultural Heritage
Architects should be deeply aware of the mission they bear, especially when designing buildings in cities or regions with profound historical legacies. How to actively care for, perpetuate, and inherit regional culture is an issue that must be dealt with.

7.6 Sustainability
7.6.1 This is not merely a slogan; it emphasizes both philosophy and technology, embodying a comprehensive architectural perspective and the DNA of architecture that transcends standard norms rather than a particular architectural style or movement. It does not solely focus on appearance and style.
7.6.2 It is necessary to consider both the present and the future throughout the entire life cycle of a project, rather than merely in the design phase.
7.6.3 Thought should always be given to leveraging the benefits of nature and making full use of all-natural resources. It refers to the use of design methods, materials, energy, and exploitable spaces that are harmless to the surrounding ecosystem or community environment, while adhering to sustainable principles of society, economy, and ecology, to reduce building energy consumption and negative impacts on the environment. Standards for green architecture include "low carbon, zero carbon, and negative carbon" levels. Specific strategies can be chosen based on the project's actual conditions, such as passive and active systems, renewable energy systems, green materials, sponge cities, and more.

7.7 Economic Viability
Economic viability underscores the cost-effectiveness of architecture, serving as the pragmatic basis for its realization. Economic viability extends beyond mere cost-saving measures, encompassing an evaluation of comprehensive considerations spanning initial investment and long-term operational efficiency. Furthermore, the rationality of architectural economics should be assessed within the context of the building's full life cycle.

7.8 Innovation
Innovation stands as an endless frontier[1] and a perennial subject that architects must confront throughout their professional careers. This statement is very profound: design is a response to social change[2] The visitor experience at the Saudi Pavilion, favored by the general public, resonates with contemporary audiences' desire for museum visits in the high-tech era. It subverts the traditional dichotomous viewing model, creating a more immersive, integrated, or perhaps more impactful visitor experience. From this perspective, creation should not focus too much energy on appearance and fashion, but on spatial experience, as fashion fades with time, while essence endures eternally.

Certainly, the criteria for excellent architecture should encompass more than just these aspects. As our research progresses, it is anticipated that additional common insights will be incorporated.

8 The "Macro Vision of Architecture" under the "*Organically-Permeated Design*"

---

[1] "Integrated and Harmonious Architecture" by Chen Xiong, published by China Architecture & Building Press, p.14.
[2] "Realize: Design Means Business" by Paul Haisch, translated by Yang Huiming, Jinghua Press, p. 38.

8.1 Objective Background

8.1.1 Entering a new era, China's development has shifted from a focus on quantity to a focus on quality.

8.1.2 While the real estate sector slows, there has been a significant increase in high-tech industrial projects and information technology (IT) projects.

8.1.3 In the era of information (or smart age), new types of buildings increasingly exhibit "intermediate characteristics" between civilian and industrial uses, and the boundary between civil and industrial use is becoming more blurred.

8.2 Subjective Background

8.2.1 Reflecting on the limitations of the binary classification system of industry and civil use, starting from the beginning of modernist architecture led by Peter Behrens.

8.2.2 Based on my affiliation with the China Electronics Engineering Design Institute (hereinafter referred to as CEEDI) — considering its, "genetic makeup", scope of business, and the total number and composition of its architectural staff — I reflect on the limitations of binary distinctions between industrial and civilian architecture.

8.3 Definition of the Macro Vision of Architecture and Its Distinction from "Broad Architecture"

8.3.1 Definition of "Broad Architecture": Broad architecture is defined as an integrative pursuit of the essence of architecture through the core role of urban design, conceptually and theoretically amalgamating the key aspects of architecture, landscape studies, and urban planning.

8.3.2 Definition of the "Macro Vision of Architecture": With the advent of the information age, traditional classification methods appear increasingly simplistic and crude. This concept seeks to abolish the traditional, simplistic division between industrial and civilian architecture, positing that any physical and spiritual space that meets human needs qualifies as architecture, termed the "Macro Vision of Architecture."

8.4 The Significance of the "Macro Vision of Architecture" for Architects

After discussing the "Macro Vision of Architecture," it is imperative for us, taking CEEDI as an example, to explore why the simplistic and deliberate bifurcation between industrial and civilian architecture is no longer tenable:

8.4.1 Tracing the development of Western architecture over the past 150 years reveals that the distinction between industrial and civilian architecture was never originally pronounced. P. Behrens, a pioneer of modern architecture, inaugurated his career with an industrial building—the AEG Turbine Factory in Berlin. His disciples, who include modernist masters such as Gropius, Mies, and Le Corbusier, also began their careers with industrial projects and seamlessly worked across both industrial and civilian architectural sectors.

8.4.2 In the contemporary era, characterized by the information age or a society inclined towards smart technologies, new architectural types are emerging, confirming the inappropriateness of a rigid classification between industrial and civilian architecture.

8.4.3 Academically, all architectural forms inherently involve process design. Facilities such as chip factories, hospitals, hotels, museums, and data centers all incorporate technical aspects, differing only in function, spatial experience, aesthetic form, and the balance required between form and function. Essentially, the nature of architectural creation remains consistent across types.

8.4.4 The current design market is characterized by an oversupply of designers relative to available projects, positioning design institutes in a reactive rather than proactive stance. The market demands that institutes maintain an open mindset, ready to tackle whatever projects arise. This market reality necessitates a flexible approach to categorization for survival.

8.4.5 Reflecting on China's shift from addressing availability ("having or not") to quality ("good or not"), clients' overall expectations for industrial buildings, especially high-tech ones, now rival those for civilian structures in terms of spatial quality and artistic value.

8.4.6 Moreover, many of our clients originally engaged in industrial production projects are now diversifying into non-industrial production ones. To continue fostering trust and extending our services, it is crucial for us to be versatile.

8.4.7 The industry's advocacy for "full-process consulting and design" and "architect-led project responsibility" calls for architects to take full responsibility for managing and overseeing the entire process across every discipline involved. For industrial design institutes, this means breaking traditional work boundaries to embrace a holistic, full-process aligned with the "Macro Vision of Architecture."

8.4.8 The historical achievements and accumulation of CEEDI, the current composition of its architectural staff, and the resulting position within China's design industry's civil architecture sector, underscore an unwarranted emphasis on the division between industrial and civilian architecture. Such a deliberate distinction undoubtedly represents a self-imposed limitation, particularly when the already modest architect workforce is further divided, weakening its overall strength. In the end, neither domain is well served by this fragmentation. Only by striking with a tightly-clenched fist can we yield greater power and effectiveness.

In conclusion: Following the architectural requirements of fundamental elements and extension is a basic professional code for architects, irrespective of the project type, which involves only different content for design. By discarding traditional binary classification and overcoming the stereotype about architectural types, CEEDI is poised to seize current market opportunities and achieve greater growth. The so-called "Macro Vision of Architecture" transcends content and typology; it is about thinking-big and mien that betray an architect's choices and values.

"Organically-Permeated Design" represents our philosophical and methodological framework, guiding our pursuit of "organic architecture." The synergy between CEEDI's "Macro Vision of Architecture," based on an in-depth understanding of architectural principles, and "Organically- Permeated Design" will undoubtedly lay a robust foundation for our architects and engineers to continue refining our existing strengths and exploring new areas in the future.

Conclusion

The Information Age has rendered access to information increasingly convenient for architects, while the advancement of AI has bolstered the potency of our modes of expression and presentation. However, the current reality reveals that the objectives of contemporary Chinese architecture are not necessarily becoming clearer or more steadfast. This circumstance underscores the significance of establishing our own ideological framework.

The theory of "*Organically-Permeated Design*" advocates for an integrated perspective to analyze and judge things, aiming to accurately grasp the internal structure and essence of phenomena. It is precisely because she emphasizes dialectical and rational thinking methods that the expression form of the work should always change due to the different natural field forces and deep structures of each project. Although architectural creation manifests as a multitude of contradictions, collisions, mergers, and re-organizations, these phenomena are not merely aesthetic and intellectual games; rather, cultural and social content should take precedence. Our exploration of architecture is not only to seek new meanings and the aesthetic possibilities of "stylisticism" but also to uncover the essential significance of architecture using this method.

After addressing the core questions of "how to perceive architecture" and "how to articulate architecture," the theory of " Organically-Permeated Design" entails substantial future endeavors. Researching and studying classical design techniques, and excellent construction details, along with innovating based on these foundations, require us to invest greater commitment and effort. It is noteworthy that even with identical principles, different methods of architectural expression yield vastly different results, and vice versa. While the perception of architecture determines the architect's depth, techniques and skills obviously belong to the second level.

Architects often enjoy discussing the rational and thoughtful aspects of their work but often overlook the flashes of passion that once inspired their designs. Undoubtedly, after establishing principles, there is still a significant amount of work to be done in methodology. Therefore, the exploration of "possibilities in technique", "search for new perspectives", and especially the "stimulation of artistic imagination" become the key to unlocking our infinite potential for innovation in the future.

As the generation that bridges the past and the future, we must remain vigilant, maintaining a balance between practice and reflection. The mission of creating contemporary Chinese architecture remains on our shoulders. Growth is impossible without a sense of urgency and crisis, which also implies that a greater crisis lies ahead.

May 30, 2024, Wuluju, Beijing

学术思考

Academic Reflection

思维结构的调整与建筑文化的新图景

王振军

引言

思维作为人类三大本质力量（即思维、体力、情感和意志）的指导机能，构成指导人类文化活动的智力。从某种意义上可以说：思维是文化的核心。所以作者认为：从思维这一文化的最深层次来剖析建筑文化的过去和将来，将会为新的建筑文化观的重构提供方法论上的意义，并为当代建筑文化的创作找出正确的出发点。

一

按思维学的定义：思维是人类以语言、符号和形象为载体抽象反映事物本质和规律的复杂生理和心理活动。而文化，从广义上讲：它是人类为适应环境，以求得群体更好地生存和发展，借助于群体内部沟通而形成的思想观念体系和共同行为模式。在这种定义下，文化就不仅仅表现在物质诸方面，更重要的还表现在精神和心理诸方面，而这方面又以思维为其核心（图1）。

建筑文化作为文化的一个重要组成部分必然受到思维这一文化核心的制约。建筑文化的主体——建筑师和社会大众因为具备了思维能力，才能正确把握客观事物运动的本质和规律，而创造出人类特有的建筑文化系统并将其记载、存储、发展。而思维的这种制约作用是通过建筑文化主体所拥有的"思维结构"的不同来实现的。

这种思维对建筑文化的制约作用，可以从中西两大建筑文化体系的比较中看出。以直觉思维为主要特征的中国传统的思维结构，反映在建筑上，即在空间构成上强调意会性和虚拟性，建筑与环境相互交融、若即若离，空间组织原则更是灵活多变；而以逻辑思维为特征的西方思维结构反映在建筑上则是建筑内外空间的明确限定，空间构成更是表现为规整严谨、条理分明。

但同时我们也应当认识到：思维与建筑文化的关系是一种相互制约、辩证统一关系，因为包括思维在内的人的各种素质是在大自然特别是社会历史中发展中产生并进化的。人类作为唯一能够积极地创造着自己生存环境的存在物，在几千年的实践活动中所创造的建筑文化反过来又在其思维结

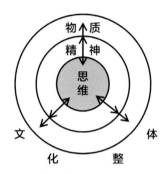

图1　文化整体的三个层次的划分

思维是其核心，具有指导性作用。但其他两个层次也对其核心有反作用。

构中留下印迹，制约着思维结构的调整。例如，某种具有典型意义的建筑（包括其形式、装饰等），一经被社会认可，就会成为一种约定俗成的符号，随着这种符号地位的日益增强，它就会反作用于文化主体，迫使他们将其作为一种新思维元素纳入其自身的思维结构中去，从而使其思维结构得以更新。

以上分析使我们看清了思维与建筑文化相互作用的关系，同时也认识到了思维的"历史性"即动态发展的这一特征。关于这一特征，我们还可以从"个体思维发生学"的角度来分析：一个人从外部实践活动到内化思维的发展和形成，大体经历了直观动作思维、直观表象思维和抽象思维三个阶段，这也就是人的概念框架即思维结构的起源和确立的过程。它是在个体思维器官发展日益成熟之后，由新的外部实践与个体原有认知图式实现协调、整合，从而内化、建立起新的思维结构。对于这一特性，发生认识论创始人皮亚查在《发生认识论原理》中做了重点阐述："事实只有被主体同化了的时候才能为主体所掌握""所谓同化是指把给定的东西整合到一个已存在的结构中或按其基本格局形成一个新结构"。由此可见，思维发展的整个过程是一个内外因相互作用的过程，是相对稳定与绝对变动的统一。

上述讨论表明，思维结构的"动态发展"特性是由思维与文化的相互作用和其自身的特点所导致。既然如此，我们就有必要把建筑文化及其主体的思维结构置于当代文化的背景下来重新进行审视，从而发掘当代建筑文化主体思维结

构调整的必要性及其趋势。

二

考查当代文化的发展状态，我们又必须从文化的最表面层次——物质层次即生产方式、科学技术、经济条件等一系列物质条件入手。受这些物质条件的制约，在古代人们只能从总体上、外观上大致地描绘自然现象，那时人类采用只是朴素的"直观总体论"的思维方式；到近代，科学刚诞生，还处于搜集整理资料、分门别类研究阶段，形成了"科学分解论"的思维结构。而紧随其后的科学三大发现（即细胞学、能量守恒与转化、生物进化论），又形成了"辩证论"的思维结构。而进入现代以后，由于相对论和量子力学等一系列新理论，特别是系统论和信息论、控制论等学科的出现，科学高度分化的同时又高度综合。科技及经济的高速发展导致了"大科学""大经济""大技术"等一些大系统客体的出现。从而引起文化载体及传播工具的高速发展。这也就是托夫勒《第三次浪潮》中所说的"信息化"社会的到来。文化物质层次的巨大变迁又反作用于它的内部层次——精神及思维，引起它们的巨大震荡。

当代文化的新图景开始出现了：生产和沟通方式的改变导致了职业分层，实践形式、家庭结构、代际关系的改变，随之社会的组织结构即文化主体的组织结构发生变化。从而引起文化的传播范围更加广阔，界限更加模糊，表现为"世界化"倾向；文化的接受方式更加多样化、高速化，在社会形态上表现为城市化倾向；文化的内容更加丰富且更加侧重于科学实证；而具体到现代文化的主体则更加表现出"拥有""综合""参与"的本质特性。现代文化以前所未有的多样性、复杂性和共享性渗透到文化的各个层面，从而对主体的思维结构构成了深重的震荡，而这种震荡也不无例外地波及建筑文化的主体——建筑师和社会大众。

思维结构的"动态发展"特性在这种震荡下，表现得更加淋漓尽致。按照信息论观点，思维系统处在不断从外界接受信息又不断向外界耗散信息的状态以维持其有序结构。新图景激发主体思维结构生发新的概念框架和方法论准则。故有的思维方法的单向性、直观性，思维品格上的实用主义和以易制难、以内驭外、以法致道、以不变应变的思维习惯等尽管也形成了光耀千秋的独特建筑体系和秩序，然而进入当代，这种传统的思维结构势必窒息了建筑文化主体全面生发和处理信息的能力，使至今我们的建筑师和社会大众都为此付出了惨重的代价。往日的理想失落，历史的经验破碎了，问题已不在于这些思维结构是否依然合理，是否曾经灵验，而在于它们已不再能够应付当代文化新图景的挑战。

这里值得指出的是：直接参与除建筑文化创作活动之外的其他文化实践活动的社会大众已率先开始调整自身的思维结构。他们的文化素养在提高，新的价值观、审美观已被不断地挖掘和展现。其个性已被最大限度地解放，大众再也不会人云亦云地评价任何建筑作品，他们以其"开放性""包容性"期待着建筑文化新图景的出现。

现实召唤着建筑师对自身的思维结构进行反思和调整，"复古主义"所造成的恶果也预示了单向、封闭的思维结构的末日。建筑文化的多样性和复杂性需要有新的思维结构去把握。这种思维结构应更加注重知识结构的更新和纵横向扩展，在建筑文化的创作中更加注重研究各种信息的系统结构、关系和整体。

三

系统论、控制论、信息论经过20世纪20～40年代的酝酿，于20世纪50年代相继诞生。我们把这"三论"运用到思维学当中，概括为"系统综合论"思维结构，它以"整体性"为基本出发点，以"综合性""动态性""信息化"为特征，要求主体在把握客体时注重以整体观点，系统观点去研究，更强调思维结构的"动态多变"特征。

新的思维结构一旦为建筑师所掌握必将引起固有建筑文化观的概念框架的错动和更选，对建筑文化观的重构有深远的现实意义。下面我们尝试顺着这一新的思维结构来勾画未来建筑文化的新图景。

（1）"整体综合性"原则：建筑师把创作客体及手段（即符号）作为一个系统整体来看待，任何系统整体都是这些或那些要素为特定目的而构成的综合体，对这些综合体的研究都必须从它的成分、层次、结构、功能、内外联系方式的立体网络作全面综合的考察，才能从多因果、多功能、多效益上把握系统整体。

近年兴起的"城市设计学""环境设计"就是系统综合思维外化的最好例证：传统狭义的设计思想已经被从大的文化结构，以满足城市居民的生理、心理要求为根本出发点，以提高城市生活的内外环境质量为最终目的，对城市的建构综合考虑的整体性创作思想所代替，20世纪60年代兴起的"后现代主义"从某种程度上反映了人类在经历了还原分析的精确思维时代之后，又开始向整体综合的系统思维体系的转换趋势（图2）。

而具体考察传统建筑文化的时候，又使我们能够去从建筑整体哲学意义上去把握，从而避免了认识上的浅肤和狭隘，使我们从以整体来反对其中一部分的自相缠绕的怪圈中解脱出来。

（2）"动态多变"原则：第三次浪潮把世界卷入了信息社会，思维结构必然在当代文化的新图景下与其他文化结构有着高频度的、立体交叉式的信息交流，从而引起固有思维结构的不断建构和破坏，呈动态发展的格局。坚持这一原则必然为建筑创作多元化思想提供内在可能性，从而使建筑主体的设计思想由传统的"以不变应万变"向"以变应变"转变。

这一准则也将改变我们对"传统"的概念，从而以动态观点来探讨"传统建筑文化"这一理论界几经论战的课题，动态观点看"传统"，应是随时间而流变，是尚未规定的东西，永远在制作之中，每个时代的建筑师都有义务对"传统建筑文化"起作用，在这一点上我国建筑已迈出令人欣喜的一步（图3、图4）。

（3）"信息化"原则：普遍说来信息在整个自然界和生物界都存在，加上当代文化传播更加开放、发达，导致各种建筑文化的交错。各种建筑文化信息在到处混合、变形，强

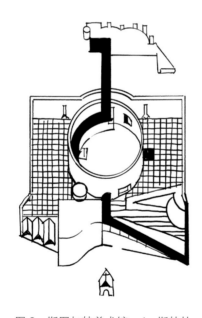

图2　斯图加特美术馆　J·斯特林

一个真正的共和国，一条曲线形步道引导，将喧闹的城市与封闭的内庭、玩世不恭的广场与中央神圣的教堂、粗琢的体量与细腻的栏杆、下沉的陶立克柱廊与橘红色金属转门，将传统与现代创造性地组合在一起——新的观念激发人们从新的角度认识建筑。

"整体综合性"原则使我们摆脱了把建筑仅作为有别于人的客观存在的局限，而建立起人类文化活动的真正内在的联系，从而促使我们从新的角度去思考建筑设计的过程；

图3　松江方塔园何陋轩　冯纪忠

传统建筑文化观中的动与静、层次、主客体有无序等概念在小的何陋轩中辗转翻腾，历史在木、竹、钢构架之间，在屋脊与檐口、墙段、护坡之间，在上下凹凸向背的转换中流变，平面中三个层次上的60°旋转使传统轴线的定向意义表现得更为精彩。

图4 松江方塔园东大门 冯纪忠

作者站在历史的高度，在建筑文化的过去、现在和未来上寻找发展契机。传统的大门被赋予新的内涵。由现代钢构架、不等坡顶和花岗岩组成的"似是而非"的传统形色所具有的深沉与含蓄传递给文化主体一个崭新的建筑意念，作者在继续制作传统。

化或减弱。建筑师面临着纵向传来的传统信息，横向的异邦信息及其他横断学科传来的信息。"信息化"要求建筑师必须将其进行最优化组合，建立起新的信息场，然后再作出选择，进行综合。符号论美学家恩斯特·卡西勒认为，"一切文化现象都是人类符号活动的结果，艺术就是运用符号来表达人类各种不同的经验""艺术的眼睛并不是被动地接受其他事物，而是一种构成活动"。日本建筑师的探索表现了他们的信息化观念，矶崎新的作品就是由综合而创造的最好例证（图5）。

"信息化"要求建筑师能够在与外界的信息交流中处于自由、开放状态，及时接收新信息，开拓视野和思路，现代设计中对"行为科学""社会学""心理学"等横向学科的引进也表明了这一信息化趋势。

当代建筑文化新图景在向我们昭示：当传统的思维结构被审视、扬弃，被朝向"系统综合论"思维结构调整后，当代建筑师所面临的思维视野实属无限。

综上所述，当代建筑文化主体的思维结构正处于激变阶段。而"系统综合论"的思维结构取向日益成为主流。当今建筑文化的丰实内涵需要去开拓。但如果我们不首先去主动寻求问题的最深层次的解决办法，那我们本身也将成为问题的一部分。

图5 筑波中心广场 矶崎新

日本筑波中心广场一堆互换了形式的实体不平衡地被放置其中。坚实之体使人联想起 Mrzio 广场的意象，而极端传统的入口蕴含于极抽象的形式之中。暗示着喧嚣与沉默的对比，这是一个能量极强的信息场，促使公众参与整合。

参考文献：

[1] 皮亚杰.《发生认识论原理》.北京：商务印书馆，2009.

[2] 张恩宏.《思维与思维方式》.哈尔滨：黑龙江科学技术出版社，1987.

[3]《世界建筑》1987年第五期.

[4] 路易·多洛.《个体文化与大众文化》.黄建华，译.上海：上海人民出版社，1987.

[5]《读书》1985年第八期.

本文引改自《南方建筑》1990年第二期

The Adjustment of Thinking Structures and New Perspective of Architectural Culture

Wang Zhenjun

INTRODUCTION

Thinking, as the guiding function of the three essential forces of human beings (i.e., thinking, physical strength, emotion and will), constitutes the intellectual that guides human cultural activities. In a sense, it can be said that thinking is the core of culture. Therefore, I believe that by analyzing the past and future of architectural culture from the deepest level of culture, which is thinking, will provide methodological significance for the reconstruction of new architectural culture and find the right starting point for the creation of contemporary architectural culture.

ONE

According to the definition of thinking, thinking is a complex physiological and psychological activity in which humans use language, symbols, and images as carriers to abstractly reflect the essence and laws of things. And culture, broadly speaking, it is the ideological system and common behavioral patterns formed by human beings in order to adapt to the environment for better survival and development of the group, with the help of communication within the group. Under this definition, culture is not only manifested in material aspects, but more importantly in spiritual and psychological aspects, and this aspect is centered on thinking.

Thinking is its core and has a guiding role. However, the other two levels also reciprocally impact its core.

As an important part of culture, architectural culture is bound to be constrained by the cultural core of thinking. The main body of architectural culture —architects and the general public—because they have the ability to think, hence grasp the nature and principles of objective things and laws of movement, and create a unique human architectural culture system and its records, storage, development. The constraining effect of thinking is realized through the different "thinking structures" owned by the subject of architectural culture.

The constraining role of thought on architectural culture, can be seen from the comparison of the two major architectural culture system of China and the West. Intuitive thinking as the main feature of the Chinese traditional thinking structure, reflected in the building, that is, the spatial composition emphasis on the intention of virtuality and sensitivity. The building and the environment are intertwined, the principle of spatial organization is flexible and changeable; however, the Western thinking structure which features logical thinking reflected in the construction of the building in the clear definition of inside and outside, the spatial composition is rigorous and well-organized.

But at the same time, we should also realize: thinking and architectural culture is a relationship of mutual constraints, dialectical unity of the relationship.Various human qualities, including thinking, are evolved in the nature, especially in the development of social history. As the only being that can actively create its own living environment, the architectural culture created by human beings in the course of thousands of years of practical activities has in turn left its imprint on the structure of thinking and constrained the adjustment of the structure of thinking. For example, a kind of typical significance of the building (including its form, decoration, etc.), once recognized by society, will become a kind of conventional symbols, with the increasing status of this symbols, it will be counteracted the main body of the culture, forcing them to incorporate it into their own thinking structure as a kind of new thinking elements, so as to make their thinking structure can be updated.

The analysis above clarifies the interactive relationship between thought and architectural culture, while also acknowledging the "historicity" of thought, that is, the dynamic development characteristic. Regarding this feature, we can also analyze it from the perspective of "individual thinking genesis": a person from the external practical activities to the development and formation of internalized thinking, roughly through the intuitive action thinking, intuitive thinking and abstract thinking three stages, which is the conceptual framework of the human thinking structure of the origin and establishment of the process. This is the process of originating and establishing the conceptual framework of human being, i.e. the structure of thinking. It is the process of internalizing and building a new structure of thinking by coordinating and integrating the new external practices with the original cognitive schema of the individual after the development of individual thinking organs becomes more and more mature. Regarding this characteristic, epistemology founder Jean Piaget emphasized in *The Principles of Genetic Epistemology*: "Facts can only be assimilated by the subject when they are integrated into the subject's cognitive structure. Assimilation refers to incorporating given information into an existing structure or forming a new structure according to its fundamental framework." From this we can see that the whole process of the development of thinking is a process of interaction between internal and external factors, a unity of relative stability and absolute change.

The above discussion shows that the "dynamic development" of the structure of thinking is caused by the interaction between thinking and culture and its own characteristics. In view of this, it is necessary for us to re-examine the thinking structure of architectural culture and its subjects in the context of contemporary culture, so as to discover the necessity of adjusting the thinking structure of the subjects of contemporary architectural culture and its trend.

TWO

In examining the state of development of contemporary culture, we must start from the most superficial level of culture again, the material level, namely, the mode of production, science and technology, economic conditions and a series of material conditions. Subject to the constraints of these material conditions, in ancient times, people could only depict natural phenomena primarily through generalized and superficial depictions, at that time, human beings used only the simple "intuitive totalism" way of thinking; to modern times, science has just been born. It is still in the stage of collecting and organizing information and researching in different categories, forming the thinking structure of "scientific decomposition theory". Followed by the three major discoveries of science (i.e., cytology, conservation and transformation of energy, biological evolution), and formed a "dialectical theory" of thinking structure. In modern times, due to a series of new theories, such as the theory of relativity and quantum mechanics, especially the emergence of system theory, information theory, cybernetics and other disciplines, science has become highly differentiated and at the same time highly integrated. The rapid development of science and technology and the economy has led to the emergence of "big science", "big economy", "big technology" and other big system objects. This has led to the rapid development of cultural vehicles and communication tools. This is the arrival of the "information society" referred to in Toffler's The Third Wave. The great changes at the material level of culture in turn affect its internal levels - the spirit and the mind - causing them to oscillate greatly.

The new perspective of contemporary culture has begun to emerge: changes in the modes of production and communication have led to a stratification of professions, changes in forms of practice, family structures, intergenerational relations, and consequently to changes in the organization of society, i.e. the organization of cultural subjects. As a result, the scope of cultural dissemination has become broader and the boundaries more blurred, manifesting the tendency of "globalization"; the way of cultural acceptance has become more diversified and high-speed, manifesting the tendency of urbanization on the social form; the content of culture has become richer and more focused on scientific empirical evidence; and the subject of modern culture has become even more manifested The subject of modern culture is characterized by the essence of "ownership", "synthesis" and "participation". Modern culture with unprecedented diversity, complexity and sharing penetrating into all levels of culture, thus constituting a deep shock to the thinking structure of the subject, and this shock is not exceptionally affected by the main body of architectural culture - architects and the general public .

The characteristic of "dynamic development" of the thinking structure is more fully manifested in this kind of shock. According to the viewpoint of information theory, the thinking system is in the state of constantly receiving information from the outside world and constantly dissipating information to the outside world in order to maintain its orderly structure. The new perspective stimulates the subject's thinking structure to generate new conceptual frameworks and methodological guidelines. Therefore, the unidirectionality of the thinking method, intuition, the pragmatism of the thinking character, and the easy to control the difficult, internal control over external factors, the law to the Tao in order to adapt to changes in the thinking habits, etc. Although the formation of a unique architectural system and the order of the light of the dry autumn, but into the contemporary era, the traditional thinking structure will inevitably suffocate the main body of the architectural culture of the comprehensive generation and processing of information, because of that, our architects and the community have paid a heavy price for this. This has made our architects and society pay a heavy price for this. Ideals of the past have been lost, historical experiences have been shattered, and the question is no longer whether these structures of thinking are still reasonable or whether they were once effective, but rather whether they are no longer able to cope with the challenges of the new perspective of contemporary culture.

It is worth pointing out here that the public, who are directly involved in other cultural practices besides architectural cultural creation activities, have taken the lead in adjusting their own thinking structures. Their cultural literacy is improving, and new values and aesthetics have been continuously explored and displayed. Their personalities have been liberated to the maximum extent, and the public no longer evaluates any architectural works in the same way as others, and they are looking forward to the emergence of a new perspective of architectural culture with their "openness" and "inclusiveness".

Reality calls architects to reflect on and adjust their own thinking structure, and the bad consequences of "retroism" also foretells the end of unidirectional and closed thinking structure. The diversity and complexity of architectural culture require a new thinking structure to grasp. This thinking structure should pay more attention to the updating of knowledge structure and vertical and horizontal expansion, and pay more attention to the study of various information system structure, relationship and the whole in the entrepreneurship of architectural culture.

THREE

System theory, cybernetics, and information theory were conceived in the 1920s and 1940s, and were born in the 1950s. We apply these "three theories" to the science of thinking, summarized as the "system synthesis theory" thinking structure, which takes "wholeness" as the basic starting point, and "synthesis", "dynamics" and "informatization" as the characteristics. It takes "wholeness" as the basic starting point and "comprehensiveness", "dynamism" and "informatization" as the characteristics, which requires the subject to focus on the overall viewpoint and systematic viewpoint when grasping the object, and emphasizes the "dynamic and changeable" structure of thinking more.

Once the new thinking structure is mastered by the architects, it will cause the conceptual framework of the old architectural culture view to be wrongly moved and selected, which has far-reaching practical significance for the reconstruction of the architectural culture view. Here we try to follow this new thinking structure to outline the new perspective of future architectural culture.

(1) The principle of "wholeness" and "synthesis": architects view the object and means of creation (i.e., symbols) as a systematic whole, and any systematic whole is a synthesis of these or those elements for a specific purpose: the study of these syntheses must be based on its composition, level, and structure. The study of these complexes must be comprehensively examined from its composition, level, structure, function, and three-dimensional network of internal and external contacts, in order to grasp the system as a whole from multi-causal, multi-functional, and multi-benefit.

The rise of "urban design" and "environmental design" in recent years is the best example of the externalization of systematic and comprehensive thinking. Traditional design thinking in the narrow sense has been replaced by holistic creative thinking which takes the physical and psychological requirements of urban residents as the fundamental starting point, improves the internal and external environmental quality of urban life as the ultimate goal, and gives comprehensive consideration to the construction of the city, and the "post-modernism" which emerged in the 1960s of this century reflects a certain extent the tendency of human beings to switch to the holistic and comprehensive systematic thinking system after the era of precise thinking of reductive analysis.

The principle of overall synthesis enables us to get rid of the limitations of architecture only as an objective existence different from human beings, and to establish the real inner connection of human cultural activities, thus prompting us to think about the process of architectural design from a new perspective; and specific examination of the traditional architectural culture, but also enables us to go to grasp the philosophical significance of the whole from the building, so as to avoid shallow and narrow understanding, freeing us from the paradoxical circle of using the whole to oppose one of its parts.

(2) The principle of "Dynamic Variability": the third wave has involved the world into the information society, and the thinking structure is bound to have a high frequency

and three-dimensional cross-fertilization of information exchanges with other cultural structures under the new perspective of contemporary culture, which leads to the continuous construction and destruction of the existing thinking structure, and presents the pattern of dynamic development. Adherence to this principle will inevitably provide an inherent possibility for the diversified ideas of architectural creation. Consequently, the fundamental design philosophy in architecture shifts from the traditional "unchanging in response to change" to a more adaptive "changing in response to change".

This criterion also transforms our understanding of "tradition", encouraging a dynamic perspective to explore the highly debated topic of "traditional architectural culture". Viewing tradition dynamically implies a fluidity with time, representing an undefined entity that is perpetually under construction. Architects from each era have a duty to influence traditional architectural culture, and in this regard, Chinese architecture has already taken a commendable step forward.

(3) The principle of "informatization": generally speaking, information exists in the entire natural and biological world, and contemporary cultural communication is more open and developed, resulting in the intermingling of various architectural cultures. Various architectural and cultural information is mixed, deformed, strengthened or weakened everywhere. Architects are confronted with traditional information coming vertically, exotic information coming horizontally, and information coming from other transversal disciplines. Informatization" requires the architect to optimally combine them, to create new fields of information, and then to make choices and synthesize them. According to the semiotic aesthetician Ernst Cassirer, "all cultural phenomena are the result of human symbolic activity, and art is the use of symbols to express the variety of human experience" and "the eye of art is not passively receptive to it things but a constitutive activity". The exploration of Japanese architects shows their informational concept. The works of Arata Isozaki are the best example of creation from synthesis.

"Informatization" requires architects to be in a free and open state in the exchange of information with the outside world, to receive new information in time and to develop their vision and thoughts. The introduction of "behavioral science", "sociology", "psychology" and other horizontal disciplines in modern design also indicates this trend of informatization.

The new perspective of contemporary architectural culture shows us that when the traditional thinking structure is examined, abandoned, and adjusted towards the thinking structure of "system synthesis theory", the thinking horizon faced by contemporary architects is infinite.

To summarize, the thinking structure of the main body of contemporary architectural culture is in the stage of radical change. The thinking structure of "system synthesis theory" is becoming more and more mainstream. The rich connotation of today's architectural culture needs to be developed. However, if we do not first take the initiative to seek the deepest solution to the problem, then we ourselves will also become part of the problem.

References:
[1] Jean Piaget. The Principles of Genetic Epistemology. Beijing: The Commercial Press, 2009.
[2] Zhang Enhong. Thinking and Ways of Thinking. Harbin: Heilongjiang Science and Technology Press, 1987.
[3] World Architecture, No. 5, 1987.
[4] Louis Dollot. Individual Culture and Mass Culture. Translated by Huang Jianhua. Shanghai: Shanghai People's Press, 1987.
[5] Reading, No. 8, 1985.

This article is adapted from *South Architecture*, Issue No. 2, 1990.

精神与形式——关于安藤忠雄与后现代主义

王振军 顾 静

众所周知,"后现代主义"是针对"现代主义"的局限性而产生的,用罗伯特·文丘里(Robert Venturi)的话来说:"它是在发现'现代主义'对当代文化的;'矛盾性'和'复杂性'状况无能为力的时候产生的"。因此,我们可以说:"后现代运动实验的最终目的是努力去使建筑显示出它应有的意义"。

但是,在当今这一"后现代"的大实验室里,欧美建筑界的一些建筑师越来越表现出他们有改变初衷的危险,他们把建筑创作当作一种时尚,把建筑的形式作为专利的工具。"复古主义"等手法的滥用也越来越表现出他们对这场"具有确定目标的运动"缺乏应有的自信。

如果说这些行动与种族的固有秉性有关的话,那么有着更深沉情感的日本建筑师在这场运动中似乎走得更坚实、更自信一些。安藤忠雄(以下简称安滕)便是其中较为杰出的一位。在他创业的短短二十多年的时间里,他以不可动摇的自信不断推出一个个富有意味的作品,从而坚实地奠定了自己在日本建筑界的地位并日益引起国际同行的关注。自1979年以来,不断获得日本建筑界的嘉奖,并于1985年获"阿尔瓦·阿尔托"奖章,他的作品也频频出现在欧美国家。而同时,随着其地位的日益国际化,他的流派归属也日益引起国际的争论。在法、德建筑界,他被看作一个后现代主义者,而在美国他则被看作一个反后现代的英雄。孰是孰非,也许还不是我们所要关注的主题,我们所关注的是:为什么安藤忠雄貌似现代派的作品却在当今的日本社会引起如此大的关注和震动?

一

安藤忠雄于1941年生于日本大阪市,他在这里长大、创业、成名,与大阪结下了不解之缘。他的家位于大阪市内一个拥挤不堪的街区里,而正是这一事实给他后来的建筑创作打下了不可磨灭的烙印,幼年时恶劣的生存经历使他从懂事的那天起就企盼着将来有一天能按自己的梦去创造一个温馨、严整的家。1969年他在自费考察欧美归来之后,在自己的家乡创立了自己的事务所。从此安藤忠雄开始了自己的建筑创作生涯。

提及自己的成长过程,安藤总是不无自豪,他说:"具备了熟练的操作技能、晦涩的理论和情感,美就无法实现"。因为他自己的思维方法不是来自训练而是来自直接的感受。也许正是这种"感受"使他的作品注入了自然、温馨、优雅、朴素、严峻、神秘等情趣。建筑作为人的思维和情感的物化,其中必然包涵着人的丰富的内心世界和深层思索。

安藤是在用作品医治着幼年时以及现今日本社会给自己心灵上的"创伤"。他说:"人类文化发展的过程中不断征服着自然的同时也征服自我生命原本的一切,它使人类在取得了巨大进步的同时也承受着生存环境的破坏、心灵的扭曲和压抑"。在介绍自己的作品时,他说:"我不是在追求表面的舒适,而是想追求随着经济迅速发展而失去的东西——与自然本质的联系、与材料的直接对话,以触发居住在生活空间中的人们的小小发现和惊奇,在简朴的生活中获得乐趣和美学意义上的高扬"。"我努力去建构生活的最基本形式,在这里居者可享受到阳光和春风"。这就是安藤的创作思想,朴素而深刻。

那么,安藤又是通过怎样的形式来表达他的创作思想的呢?据他自己说:在创作房子的时候我总是在思考着约瑟夫·奥伯(Josef Albers)"对方形的敬意"和乔凡尼·巴蒂斯塔·普林西(Giovanni Battista Piranesi)的"Carceri(描绘一个迷乱空间)"这两幅画。前者抽象,表现了安祥与温馨,后者具有强烈的"表现主义"色彩,动荡不安,这两幅作品是那样的令安藤迷恋,难以忘怀,以至于几乎在他的所有作品里都有这两幅画的影子。他用自己偏爱的圆和方,创作着有序的建筑空间,建筑成为根据严格几何关系建构的实体加上他所偏爱的素面混凝土,乍看上去颇有柯布西耶的遗风,这也就难怪欧美建筑界对其议论纷纷了。然而,他的内部空间是经过了精心转换的,洋溢着特殊的韵味和品质,也许这就是克里斯蒂安·诺伯格-舒尔茨(Christian Norberg-Schulz)所追求的"场所精神",这种"精神"是无法用照片来捕捉的,它需要人在动态过程中去把握、去体验,所以可以说安藤的建筑实体是承载着人类印迹的实体,而实现这种转换的具体手法是:赋予作品以"迷宫性"(Labyrithine Quality),而这一手法一般都是在"联结体"处使用。安藤在谈到他的创作思想时又补充道:为了弥补人类理性所造成的缺陷,我又引入了"自然"这一因素,它与"迷宫性"一起构成了建筑的"表现特性",而这一特性正是勒·柯布西耶(Le Corbusier)所欠缺的。

但是观察安藤的作品就可以确认,他从现代派大师那里继承了不少的东西。他从勒·柯布西耶(Le Corbusier)那里继承了"立体主义"的构图法则,从路易斯·康(Louis Kahn)那里继承了对材料本身的偏爱和对"光"执着的使用。提到路易斯·康(Louis Kahn),可以说在西方现代派大师中,他是对包括安藤在内的日本第四代建筑师造成影响最大的一位。

二

Kidosaki住宅是安藤经过六年的反复设计于20世纪70年代建成的。业主在建设过程中聘请丹下健三出任总建筑师,在丹下健三这位日本建筑界以严格和宽厚而著称的老师的诱导下,安藤按自己志愿创造了杰作。

这幢住宅是为一对夫妇及其双方父母而建造的,这合乎礼仪的要求却带来了复杂的空间问题。但是在安藤朴素、严谨的构思下,每个部分都处理得很完整,空间内被赋予一种"被净化的平衡"(Poise and equilibrium)。内部没有用随意的暗示去渲染。空间流动的强力促使人们去经历、去体验,很难用图片来说明,因为他不是那种程序化、概念化(如勒·柯布西耶(Le Corbusier)的"五原则")操作的结果,而是来自内心情感及思想的创造。这其实也就是克里斯蒂安·诺伯格-舒尔茨(Christian Norberg-Schulz)所说的"存在主义建筑"(Existen-tial Architecture),也就是说,这种建筑只有通过亲身经历才能理解。均衡的比例和虚幻的光线被注入里面,柔和的"灰"唤起人们内心深处一种无法描述的感受,日本人叫做"幽玄"——一种深远、神秘的感受,内部充满一种伟大的意象。

由于日本土地私有制的原因,许多分块小且开发价值不大的土地至今仍保持着原状。这样,大阪市内就出现不少木结构住宅、"简易建筑"和后来的钢筋混凝土新建筑混杂分布的地段。继Kidosaki住宅之后,安藤在以后的二十多年里,在这种基地上陆续完成了一系列作品,"住吉长屋"(New Uersion of the Old Row House),也是其中之一,且使他一举成名,荣获1979年"日本建筑协会年度奖"。

这是位于大阪的一块狭长用地。安藤为了在上述杂乱的环境中创造出既保证私密性,又接触大自然的居住空间,采用了立面完全封闭的混凝土盒子形式,中央三分之一处设采光天井,其中设外楼梯和天桥来联系上下左右。房间通过天井接触大自然的阳光、空气、风雨。这富有平静感与遮蔽感的天井把人们从周围嘈杂的环境引向天空。在构图上,安藤采用了"剪裁"与"并置"的手法,联结两个体量的节点是天井和楼梯。通过对安藤以上两个早期作品的体验,我们可以感觉到:与外界的沟通与隔断的矛盾冲突在他早期作品中体现得淋漓尽致,幼年留下的创伤仍未愈合。他在小心地回避着、谨慎地选择着纷杂的世界。而回避的方法也是很诚实的:灰色、厚重的混凝土墙,墙上偶尔开出狭长的镂空。而房子的内部则是一个逐渐开放的系统,这正是安藤内心世界的写照。"外面的世界很精彩,外面的世界也同样无奈"。这种矛盾心理在他的创作早期一直没有得到疏解。在特定的基地里,他一丝不苟地营造着自己的房子,丰富着自己的"空间美学",实现着自己幼年时就已勾划好的"理想世界"。安藤在每一块特定的基地上"自治"自己的天地,他早期作品所显示出的空间效应、形式及其与城市环境的对立和隔断,很明显地限制了他的发展。

尽管如此,他的早期作品已显现出一种特殊的魅力。因为他是在全力以赴地用自己的情感去试图创造一种"共同的符号秩序"来作为对自然、对周围世界的一种反应。但他的这套"符号秩序"既不同于罗伯特·文丘里(Robert Venturi)的"习惯元素",也不同于摩尔的"片断联想"。他试图寻找出建筑的"原型种类",像路易斯·康(Louis Kahn)那样。

三

如果说在早期的"住宅"设计中,安藤通过自己的直觉、情感较为圆满地渲染了"世俗世界"的话,那么在以后的一系列教堂设计中,他又成功地塑造了"神圣世界"。风之教堂(六甲山教堂)坐落在一个山的山顶,建筑由两个部分组成：入口柱廊及教堂大厅，柱廊由 3m 宽的柱子支撑，近 40m 长的空间似乎在无限地延伸，安藤在很有耐心地培养那些朝拜者虔诚的心，这是通向教堂的必由之路。进入其中，你会发现自己被柔和的光包围着。由于柱廊的穹顶由毛玻璃覆盖，人仿佛走在苍穹之下，一切都不再神秘，一切都不再具有实质。匍匐在苍穹下，耳边是风的哨声和树的婆娑声，丰足的光造成的"缥缈感"加上"只闻自然而不见自然"的事实使人感觉正行进在通往另一世界的通道上——通向一个神灵的世界。在通道的尽端，下去几步，经一片曲墙引导，人们终于发现了另外一个世界。这里很幽暗，一个简洁的钢十字架高高悬挂在正墙上，一束顶光由上倾泻下来。教堂的旁边是一片很缓的绿坡地，整齐得几乎看不到一点杂草，而且四周由墙围合着，安藤把它叫作"过渡区"——既开放，又封闭，介于室内与室外之间，它的确是自然的，但又经过了精心的修整和围合，一种对自然的含蓄引入。安藤这种处理自然的态度和手法贯穿在他创作过程的始终。安藤的另一个宗教作品是 1985 年设计的"水上教堂"，它坐落在日本北海道的一块平原地带。基地对面有一个人工湖。安藤首先用一片围墙将基地界定了起来。人们顺着这片矮墙，沿着一个缓坡，被引入到教堂的门厅，门厅由玻璃围合，这是一个光的世界。四个十字架四臂交融门于苍穹之下，人们沐浴在自然光之中，光的倾泻创造了庄严的气氛。在环绕十字架一周之后，顺着一个幽暗的弧形楼梯下来。突然间，一汪碧水映入眼帘，一尊十字架立于湖中，一个混凝土框把苍天与大地分开，把理想世界和世俗分开。

这一经过精心组织的景观在外观上随着季节的变化而变化。在这种转换中，人们可以感受到一种自然而神圣的精神的存在。岁月流逝，这光、这水、这天空又在呼唤另一种曲调。在经历了安藤的两个教堂之后，我们可以看出安藤通过理性的手法同样创作出了具有强烈感情色彩的作品。

四

安藤在用自己的作品解释着"人""神""自然"和"空间"。一座建立在人工岛上的办公性建筑，"方正"与"迷宫"这两幅画继续在这个设计中起着控制作用。与前期设计手法有所不同的是，安藤在经营内部空间的同时更加注重与外部既有环境的沟通。日本的"灰"空间被用来一步步软化建筑与基地的关系，丰富的层次感使这幢平凡的办公建筑变得富有生机和韵味。若是请密斯来给这幢房子下评语的话，他肯定会说安藤太拖泥带水甚至是自作多情。然而在当今多元化的社会？即使是在当时"现代主义"风行的时间里，人们的日常生活也并不是突然变得"理性化"和"实用主义"，也仍然有一个"质"的世界存在，在这之中人们具有的是一种诗化的关系而不是"理化"的关系。安藤就是试图在这种办公建筑中加入"诗化的语言"。

玫瑰花园是安藤在日本神户一片高级住宅区中设计的一幢商业建筑，这片住宅是明治时代的欧式深房，以红砖砌筑。为了与周围环境协调安藤舍弃了自己酷爱的混凝土。在设计过程中安藤以他坚强的意志，克服种种困难，设计出了顾客和业主喜爱的商业建筑。他利用地形的高差，在这建筑中设置了一个很欧化的小广场，这里阳光充裕，人们下到半层又来到另一个商店，"并置"的手法再次运用，只是比以前更娴熟。

五

安藤忠雄给我们的启示：

1. 有意义的形式并不是来自于他的结构、材料或其他，而是来源于"想象"的创造。"形式主义"的作品将是没有生命力的。至今风行的"高技派"事实上是乌突邦式的"现代主义"的最后骚动。他们只是以人的日常生活作为其出发点，这也就是他们不能永久的根本原因所在。建筑师的创作要有坚实的根基。他们的作品要能够与环境建立起有意义的联系，能够表现人类生存的基本性质，而不是其他什么表面的东西。有了这样的认识高度，就可以避免在创作中盲目抄袭别人、展示花拳绣腿等弊病，从而使每件作品都有自己的特点和深度。

2. 欲使我们的建筑独树于世界之林，最主要的不是与别人攀比经济上的宽裕、材料上的高级等客观条件，而是应当把自己具有一定思想深度的追求贯彻到自己的创作中去，使之具有独特的品质。安藤的追求就是一个很好的范例。

3. 中国有句话：和实生物，同则不继。(《国语·郑语》）——事物只能在对立统一中发展，相同的元素组合在一起则不能发展。因此，从宏观角度看，在后现代的大实验中，尽管安藤走得更扎实一些，但同时对那些欧美建筑师应当宽容一些。也许，这种追求方法的不同正可创造出各种各样的思想，形成一种动态平衡，正好可以在相生相克中防止这场运动的畸形化，从而推动人类的建筑文化朝着正确的方向发展。

在茫茫宇宙中，人无论多么伟大，他所发现的真理都是相对于他的视角而言，宇宙中还有无限的视角等待人们去发现真理。因此一种思想不应否定另一种思想，而只能互补，相反的事物往往相成。这也许正是我们今天新一代建筑师亟待形成的一种崭新的思维方式。而这也正是笔者写作此文的初衷所在。

本文引改自《时代建筑》1991年第1期

Spirit and Form: Tadao Ando and Postmodernism

Wang Zhenjun Gu Jing

It is well known that "Postmodernism" arose as a reaction to the limitations of "Modernism". As Robert Venturi eloquently noted, it emerged when Modernism's incapacity to address the "contradictions" and "complexities" of contemporary culture became apparent. Thus, the ultimate goal of the Postmodern movement has been to strive to endow architecture with its deserved meaning.

However, in today's "Postmodern" laboratory, some architects in the West increasingly show a perilous shift from their original intentions, treating architectural creation as a mere trend and using form as a proprietary tool. The misuse of "Revivalism" and other tactics increasingly indicates their lack of confidence in this purpose-driven movement.

If these actions relate to inherent racial characteristics, then Japanese architects, with their deeper emotional connections, seem to navigate this movement with more solidity and confidence. Tadao Ando stands out among them. In the short span of his career of over two decades, he has continually introduced meaningful works with unshakable confidence, thus firmly establishing his status in the Japanese architectural community and increasingly attracting the attention of his international peers. Since 1979, he has received numerous accolades from the Japanese architectural community and won the Alvar Aalto Medal in 1985. His works have been frequently exhibited in Europe and America. Meanwhile, as his status becomes increasingly international, the debate over his stylistic affiliations intensifies. In the architectural circles of France and Germany, he is considered a Postmodernist, while in the United States, he is seen as an anti-Postmodern hero. The true classification may not be our primary concern; rather, we focus on why Tadao Ando's ostensibly modernist works have stirred such significant attention and impact in contemporary Japanese society.

One

Tadao Ando was born in 1941 in Osaka, Japan, where he grew up, launched his career, and achieved fame, forging an indelible connection with the city. His childhood home, located in a densely populated district of Osaka, left an indelible mark on his architectural creations. The harsh living conditions of his youth instilled in him a lifelong aspiration to create a warm, orderly home of his own design. After self-financed study tours of Europe and America in 1969, Ando established his architectural practice in his hometown, marking the beginning of his illustrious career in architecture.

Ando often speaks proudly of his formative years, emphasizing that beauty cannot be achieved merely through skilled technique and obscure theories and emotions. His method of thinking stems not from formal training but from direct sensory experience. It is perhaps this reliance on "feeling" that infuses his works with qualities such as naturalness, warmth, elegance, simplicity, severity, and mystery. Architecture, as a materialization of human thought and emotion, inevitably encompasses the rich inner world and deep reflections of its creator.

Ando uses his work to heal the emotional "scars" inflicted by his childhood and contemporary Japanese society. He observes that while human cultural development has continually conquered nature, it has also conquered the very essence of human life, leading to tremendous progress alongside environmental destruction and psychological distortion. In discussing his work, Ando states that he does not seek superficial comfort but rather aims to reclaim what has been lost with rapid economic development: a connection with the natural essence, direct interaction with materials, and the small discoveries and surprises that enhance the lives of those dwelling within his spaces. He strives to construct the fundamental forms of living, where inhabitants can enjoy sunlight and the spring breeze. This is the essence of Ando's creative philosophy—simple yet profound.

How does Ando express this philosophy through his forms? According to him, when designing houses, he often contemplates Josef Albers' "Homage to the Square" and Giovanni Battista Piranesi's "Carceri" series, which depict disorienting, maze-like spaces. The former is abstract, evoking tranquility and warmth, while the latter possesses a turbulent, expressionistic quality. These artworks deeply influence Ando, leaving their imprint on nearly all his works. He prefers to use circles and squares to create ordered architectural spaces, building structures based on strict geometric relations and his favored material, fair-faced concrete, which at first glance may seem reminiscent of Le Corbusier. This affinity might explain the varied discourse on his work in Western architectural circles. Internally, his spaces are carefully transformed, exuding a special flavor and quality that perhaps align with what Norberg-Schulz termed the "spirit of place". This "spirit" is elusive, not easily captured in photographs but experienced dynamically by those interacting with the space. Thus, Ando's architectural forms are laden with human impressions, achieved through specific techniques that impart a "labyrinthine quality" to the works, often employed at the junctures of structures. To compensate for the deficiencies caused by human rationality, Ando also integrates "nature" into his design, combining it with the "labyrinthine" to form the "expressive characteristics" of architecture, characteristics that Le Corbusier might have lacked.

Yet, a closer examination of Ando's oeuvre confirms that he has inherited much from the modernist masters. He adopted the Cubist compositional rules from Le Corbusier and a fascination with material integrity and the meticulous use of "light" from Louis Kahn. Speaking of Kahn, it can be said that among the Western modernist masters, he has been the most influential on the Japanese fourth-generation architects, including Ando.

TWO

The Kidosaki House, constructed in the 1970s after six years of iterative design, stands as a testament to Tadao Ando's architectural genius under the mentorship of Kenzo Tange, who served as the chief architect during its construction. Known for his rigor and magnanimity, Tange guided Ando in fulfilling his creative ambitions, culminating in this architectural masterpiece.

Designed for a couple and their parents from both sides, the house addresses complex spatial issues while adhering to traditional Japanese etiquette. Ando's minimalist and strict design approach ensured that each space was meticulously crafted, achieving "Poise and Equilibrium". The internal spaces of the house do not rely on arbitrary decorative cues but rather facilitate a powerful flow of movement that encourages experiential interaction, challenging to capture in photographs. This design outcome stems not from procedural or conceptual methodologies, like Le Corbusier's "Five Points of Architecture", but from a profound engagement with emotional and intellectual creativity. This approach aligns with what Norberg-Schulz described as "Existential Architecture", which asserts that such architecture can only be fully understood through personal experience. The residence incorporates balanced proportions and ethereal lighting, invoking a profound, indescribable feeling known in Japanese as "yugen"—a deep and mysterious sensation. The interior is imbued with grand imagery.

Due to Japan's private land ownership system, many small and economically insignificant parcels remain undeveloped. This results in a mix of pre-war wooden houses, post-war makeshift constructions, and more recent reinforced concrete buildings within the urban landscape. Following the Kidosaki House, Ando completed several similar projects over the next two decades on such sites, including the renowned "Sumiyoshi Row House", which brought him fame and the 1979 Architectural Institute of Japan Annual Award.

Located on a narrow site in Osaka, the Sumiyoshi Row House was designed amidst this cluttered environment to create a private yet nature-connected living space. Ando opted for a completely enclosed concrete box design with a central skylight that houses external staircases and bridges connecting various levels. The rooms engage with the natural elements—sunlight, air, and rain—through the courtyard, which simultaneously shields the inhabitants from the noisy surroundings and guides their gaze towards the sky. In terms of composition, Ando employed techniques of "cutting" and "juxtaposition", with the junctions of two volumes being the courtyard and staircases. These early works vividly reflect the conflict between connection with and separation from the outside world, showing that the scars of his childhood had not yet healed. Ando cautiously navigates the complex world, using thick, gray concrete walls with occasional narrow openings. The interior of the house unfolds as a progressively open system, mirroring Ando's inner world: "The outside world is fascinating, yet it is also filled with despair". This contradictory sentiment persisted in his early creative phase. In specific sites, he meticulously crafted his homes, enriching his "spatial aesthetics" and realizing the "ideal world" he envisioned from childhood. Ando "autonomously" established his domain in each specific site, and his early works demonstrate significant spatial effects and forms, as well as a stark contrast and separation from the urban environment, which notably constrained his development. Despite these constraints, Ando's early works exhibited a unique allure. He passionately used his emotions to attempt to create a "common symbolic order" as a response to nature and the surrounding world. Yet, his "symbolic order" diverged from Robert Venturi's "familiar elements" and Charles Moore's "fragmented associations", as he sought to discover architectural "archetypes" in the manner of Louis Kahn.

THREE

In his early residential designs, Tadao Ando successfully captured the essence of the secular world through intuitive and emotional renderings. As his architectural journey progressed, he ventured into a series of church designs, masterfully sculpting the realm of the sacred. The Church of the Wind, perched atop a mountain, comprises two main elements: an entrance colonnade and the main chapel hall. The colonnade, supported by three-meter-wide pillars and stretching nearly 40 meters in length, seems to extend infinitely, patiently nurturing the devout spirits of its pilgrims as they make their way toward the sacred hall. Upon entering, one is enveloped in a soft light. The vault of the colonnade, covered with frosted glass, creates the illusion of walking under the heavens—demystifying the spiritual and dissolving the material. Crawling under the celestial dome, accompanied by the whistling of the wind and the rustling of trees, the ethereal light and the exclusion of direct natural views create a sense of progression toward another world—a realm of the divine. At the end of this pathway, a few steps down lead through a curved wall, guiding visitors to discover another world. This space is dimly lit, with a simple steel cross suspended high on the main wall and a beam of light pouring down from above. Adjacent to the chapel lies a gently sloping green lawn, meticulously manicured and nearly devoid of weeds, enclosed by walls. Ando refers to this as a "transitional area"—both open and enclosed, straddling the indoor and outdoor realms. It is undeniably natural, yet carefully curated and bounded, a subtle introduction to nature. This method of engaging with nature permeates Ando's entire creative process. Another of Ando's religious creations is the "Church on the Water", designed in 1985, located on a plain in Hokkaido, Japan, facing a man-made lake. Ando began by defining the site with a perimeter wall. Visitors are led along this low wall down a gentle slope into the church's foyer, enclosed by glass—a realm of light. Underneath a cross with arms that seem to merge with the sky, people are immersed in natural light, which creates a solemn atmosphere. After circling around the cross, they descend a dark, curved staircase. Suddenly, a pool of azure water comes into view—a cross stands in the lake, framed by concrete that separates the heavens from the earth, the sacred from the secular, separating the ideal world from the mundane.

This carefully orchestrated landscape transforms visually with the seasons, allowing visitors to experience a presence that is both natural and sacred. As time passes, the light, the water, and the sky call out in a different tune. Through experiencing Ando's two church designs, it becomes evident that Ando, through rational techniques, also creates works infused with intense emotional resonance.

FOUR

Tadao Ando elucidates themes of the human, divine, natural, and spatial through his architectural works, exemplified by an office structure on an artificial island. This building showcases persistent motifs of "formality" and "labyrinth", maintaining a commanding influence throughout the design process. Departing from his earlier techniques, Ando now focuses more intently on the nuanced management of internal spaces and fosters a deeper dialogue with the external environment. The strategic use of Japan's "grey space" gradually softens the interface between the architecture and its setting, lending vitality and layered complexity to an otherwise ordinary office building. If Mies van der Rohe were to critique this structure, he might find Ando's approach excessively sentimental. However, in a diverse contemporary society—even at the peak of Modernism—daily life did not wholly convert to rationalism and utilitarianism; instead, a qualitative world persisted, defined by poetic rather than rational relationships. Ando aims to infuse this office building with a "poetic language", enhancing the poetic interplay between form and environment.

In Kobe's upscale residential district, which features Meiji-era European-style mansions built from red brick, Tadao Ando designed the Rose Garden, a commercial building.

In alignment with the surrounding context, Ando abandoned his favored material, concrete. Throughout the design process, he faced and overcame numerous challenges with his resilient will, resulting in a commercial structure that is cherished by both customers and owners. Capitalizing on the terrain's elevation differences, Ando created a distinctly European plaza within the building. This sunlit plaza connects people to a lower semi-level leading to another store, demonstrating Ando's sophisticated use of "juxtaposition", a technique that he applies with greater mastery than before.

FIVE

The inspiration given to us by Tadao Ando:

1. The significance of architectural form does not merely stem from its structure, materials, or other tangible elements, but rather from the creative force of imagination. Works rooted solely in formalism lack vitality, and the popular "High-Tech" movement represents, in essence, a utopian form of Modernism's final upheaval. These designs originate primarily from everyday human activities, which partially explains their inability to establish permanence. Architects must build upon a solid foundation; their creations should forge meaningful connections with the environment and embody the fundamental qualities of human existence, not merely superficial attributes. With such an understanding, architects can avoid the pitfalls of blindly copying others or engaging in superficial design, thus ensuring each project possesses unique characteristics and depth.

2. For our architecture to stand distinctively in the global landscape, the focus should not be on competing over economic wealth or superior materials, but on integrating a depth of thought into our creative endeavors, thereby imbuing our work with unique qualities. Tadao Ando's approach serves as an exemplary model of this pursuit.

3. A Chinese proverb states: "Among similar entities, there is no succession."(*Country Language- Zheng Language*) This highlights that development occurs through the dialectical unity of opposites; homogeneous elements together do not progress. Thus, in the grand experiment of postmodernism, while Ando's approach may be more grounded, we might also afford some leniency towards Western architects. Differences in pursuit methods can generate diverse ideologies and create a dynamic equilibrium, ideally preventing any deformations in the movement and pushing architectural culture towards the correct development trajectory.

In the vast universe, no matter how great a human may seem, the truths they uncover are relative to their perspectives. The universe holds infinite perspectives waiting to be explored. Thus, one ideology should not negate another; rather, they should complement each other. Often, opposing elements are interdependent. This, perhaps, represents a novel mode of thinking that the new generation of architects urgently needs to develop. This notion also forms the initial intent behind this article.

This article is adapted from *Time + Architecture*, Issue No. 1, 1991.

轴线手法与建筑空间及形式的层次感塑造

王振军

形式的创造并不是什么神秘的能力，有许多途径可以帮助人们去开发。本人认为在各种创造形式的途径中，轴线手法是相当有潜能的途径之一。

众所周知，近年来在国际范围内，建筑上的古典主义主要在两个方面发展：一是装饰细部，二是构图法则。而后现代主义中的古典主义企图复活古典主义的装饰方面，但本人认为正是那些构图法则对整个人类的建筑景观产生了最根本、最显著的影响。同时研究轴线手法这一中国传统建筑的经典构图法则，又具有"保持建筑文化连续性"的现实意义，有人形象地把它比做接线工作。

空间秩序和层次的建立是人类空间体验的本能需求，它对空间的使用和识别都具有重要的意义。"优秀的建筑中，结构的、功能的和审美的序列必须是有机而紧凑的，建筑的首要任务就是要以这种组织程序的方法进行设计。"中国传统建筑即十分重视层次的塑造，常常通过照壁、屏风来组织室内外的空间序列，同时在组群中注重通过主次轴的引导和转换来追求层次趣味。本文拟从界定轴线元素的转换和轴线本身的转换两个方面来探讨形式和空间的层次感的塑造问题。

一 界定轴线元素的转换与层次感塑造

这种转换可分为沿轴线方向的纵向转换和相对于轴线的横向转换。

1. 纵向转换——脱开和分离

我们在分析具体方法之前有必要首先分析一下层次感的视觉及心理基础。鲁道夫·阿恩海姆（Rudolf Arnheim）对图形层次的研究是在分析"图-底"之间的转换关系的基础上展开的，他指出：形式构成不仅要注意形式（图）从背景（底）中分离出来的诸种条件和限制，而且也应该注意"底"本身在形式构成中的重要作用。层次的形成是图-底相互作用的结果，因此形成层次的图底即轴线纵向上界定元素之间的相互关系和联系程度构成了纵向序列中不可忽视的因素。其中脱开和分离不失为一个好方法。

这种方法应用到实际创作中往往激发出建筑师的创造性。如在合肥蜀山烈士纪念碑设计中，作者在总体布局上运用了轴线法，且对界定轴线的元素纪念碑碑体进行纵向的脱开和分离处理（图1），既突破了常规，又很好地塑造了纪念性空间的序列。对纪念碑建筑传统的"叙事诗"式的平面布置进行转换，这种不落俗套的处理使之在竞赛中夺魁。迈克尔·格雷夫斯（Michel Graves）的汉索曼住宅（Hanselman House1967）同样运用了轴线上元素的脱开处理，既丰富了住宅的层次，又构成了令人耳目一新的住宅形象。"住宅像电影般层层化出，人们可以在这两个层次之间细心玩味，品尝归来时心头所充满的家庭的甜蜜（图2）。"

从以上实例分析可看出，处理元素图底之间的关系成为层次感塑造的关键，以上作品的成功之处在于：

（1）利用了轴线的导向和组织秩序的作用，强化了元素间的历属关系，从而在此基础上来转换，否则容易造成形式层次与空间层次的脱节。

（2）界定元素的纵向脱开处理在空间维和时间维均增加了层次，因此更丰富了轴线在功能上的意义。如纪念碑之间的轴线能促使观众联想和遐思，住宅中的桥成为联系主体部分和工作室的通道。

（3）元素间在形式及空间上的呼应。如格雷夫斯的汉索曼住宅，主体上虚角部分暗示着入口部分的脱出。

2. 横向转换——断开

主要是指与轴线垂直的向度上界定元素在形式和空间上的转换。如同语言学中相同语法结构下单词位置的不同可引起句子意义变化一样，界定元素在轴线上的横向变异又会产生另一种效果。

横向脱开又叫断裂。日本建筑师东孝光把它叫作"间隙"（Slit）。总之，它可以看作是将一个整体分裂之后形成的二元对立这样一种方法和状态。而断开后所形成的轴向空间充当了"调和因素"的角色，它的空间特征不是单纯的，而是并置、渗透等连接方法同时使用所构成的。正如同时性的声音的引入可无限地丰富音乐的结构一样，轴向空间使各种不同的元素有可能同时表现，而不会像古典构图中的对称那样，由于轴向空间的连接而使双方的特征同时消失。

从功能上讲，横向断开处理，在内部进一步完善了整体的空间秩序，有利于功能分区，同时使内与外自然连续。

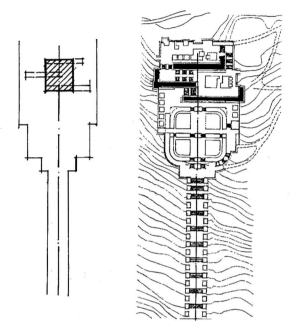

图 1 合肥蜀山烈士纪念碑

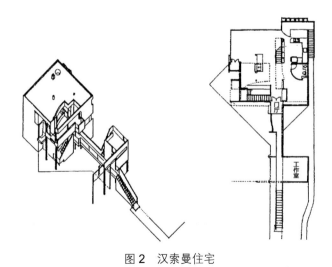

图 2 汉索曼住宅

在1982年巴黎新歌剧院国际设计竞赛中，中国香港建筑师严迅奇就采用了"断开"的处理方法。从巴士底广场中的七月柱引出的一条轴线把基地断裂为两部分：右边是主体部分，采用对称形式以保持传统城市的边界特征；左边为辅助部分，采用自由形态。作者利用间隙作为一条商业街。"这条新街是完全步行的，因而可举行各种活动，使巴黎街道所具有的著名的人情味能渗透到整个环境中，进而渗透在剧院的各项演出活动中，鼓励更多的公众参与"（图3）。

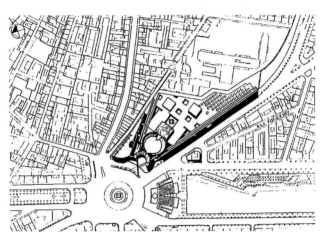

图 3 巴黎新歌剧院国际竞赛入选方案

另外，在断开处进一步进行处理，还可造就更丰富的空间层次和立面效果，如将结构、围栏、交通等形式构成的各方面矛盾予以大胆暴露，就是一种有效的处理方法。东孝光在其"原野之家"住宅设计中将裂缝空间里的二层栏板扭曲后暴露了出来，使间隙的动感和层次感加强（图4）。弗兰克·盖里（Frank Gehry）在某法院教学楼中的轴线上设置裂缝，而为了强调入口，又在裂缝处伸出楼梯并加以扭转，从而丰富了立面效果（图5）。

图 4 "原野之家"住宅　　图 5 美国某法院教学楼

3. 纵横向同时转换

上面我们对界定元素相对于轴线的两个方向的转换进行了探讨，而在实际创作中这种转换往往是同时进行的。

日本建筑师 Katsuhiro Kobayshi 对这种方法进行了研究。图6是他对其作品所做的构成分析。

第一步，确定轴线。整个群体经分析后被划分为若干个几何体量，并考虑周围环境脉络以确定轴线。

第二步，按古典构图的手法如向心性、对称性来规整各个部分。

第三步，将于轴线有界定作用的元素进行纵横二个向度的脱开、分离处理，同时将元素进一步处理成自由的几何形态，从而赋予一种偏离感。

第四步，反映在立面上的各个元素被依据每个方案的性质实行进一步刻画。

在秩序感极强的构架中，引入异质因素（如片断的无序和差异），对改变作品的静态和冷漠的规整感，体现个性和时代性，将是一种有效的手段。华盛顿罗斯福纪念碑就是运用该方法取得成功的例子，建筑师将轴线所具有的纪念性、方向感和对秩序的组织能力与曲墙的不定性、飘离感巧妙地揉合在一起，刻有文字的曲墙被相对轴线扭转成不同的角度，为参观者提供了多角度的视域和更丰富的信息量（图7）。

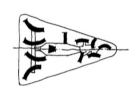

图7 华盛顿罗斯福总统纪念碑

4. 界定元素与轴线的相对旋转

将界定元素和轴线各自都保持自身的状态，然后使轴线非直角地引入到界定元素中，也可塑造出富有层次感和识别性的作品来。同时这种成非直角的介入比直角介入往往更能为形式的构成带来新意和活力。当然这种相对旋转应是有根据的，应在分析原有结构后进行。

贝聿铭在普林斯顿大学的学生宿舍设计中根据学生行为的便捷性特点和地形特征（原来就有一条斜道联系东北角的火车站和体育馆），确定了一条公用步行道，与整个校区成45°。这一方面维持了学生原有的行为心理约定，另一方面为整个宿舍群设计带来活力（图8）。

二 轴线本身的转换与层次感的塑造

城市、组群及单体建筑之所以能呈现出多样的形态和层次，除了界定元素自身可有多种表现外，组织界定元素的轴线本身也可通过一定的转换和组合表现出多样性，因为它的

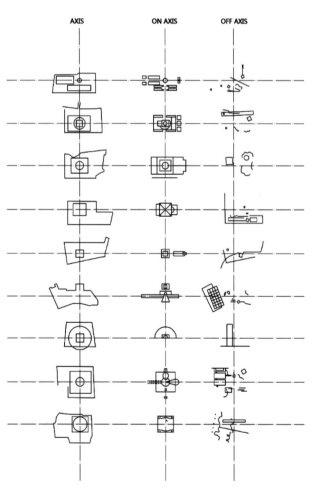

图6 Katsuhiro Kobayshi 对自己作品的构成分析

图 9 为山西五台山某寺院的入口处理。受地形和风水观的影响，无法正对闹市设门，匠人由此对入口序列做了转折处理。有时为适应山势，匠人们常常通过几段轴线来组织组群，轴线顺地形曲折蜿蜒，在塑造庄严肃穆气氛的同时，又渲染了寺观超尘拔世的浪漫。人们熟知的苏州园林更是将以上手法运用得淋漓尽致。

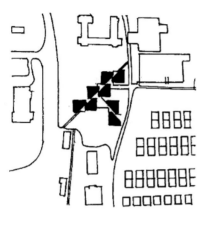

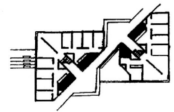

图 8 普林斯顿大学学生宿舍

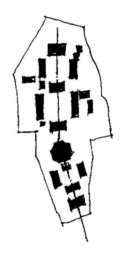

图 9 山西五台山某寺院

转换导致了空间方向上的和节奏上的根本变化。

就轴线本身转换而言，不外乎几何和拓扑两种组合关系。这一线性向度上的组合，主要是指为了空间层次塑造的需要及其他因素，如地形限制、功能要求时，而将几段轴线组合在一起。一般来说，单一轴线作为一种线性媒介一直贯穿始终，人的情绪仅仅是一种递进和量的积蓄，而通过轴线本身转换可打破单一轴线平直单调，以转折来活跃人的视线，增加景深以追求质的变化。而以上所说的方法在中国古代的山林古寺及私家园林设计中均有所应用。

1. 古代山林古寺及私家园林的启示

受社会文化因素的制约，中国传统建筑组群从总体上形成了中轴一贯、左右对称的格局。但是古代匠人在长期的实践中，通过对轴线特性的领悟，使自己从泥古不化的僵化图形中解放出来，从而能有许多不同气候、地形、类型条件下，采用轴线的偏折、错位、迂回等灵活手法，这样在当时的条件下，一方面使工程建设合乎实际条件而经济便利，同时也维持了人们的社会文化约定。这种情况在山林古寺、私家园林中出现较多。

2. 轴线的拓扑性转换

拓扑关系是指一种没有严格的数和几何性的要素间的关系，具体到轴线上则是指：没有确定角度关系的线性组合关系。虽然拓扑关系来源于一种质朴的对环境的体验，是一种对"无序"的低层次、直观本能的处理，但是进入当代，建筑师将它有意识地运用到创作中，常常会给形式构成带来新意。

查尔斯·摩尔（Charels Moore）的克累斯格（Kresge）学院宿舍区设计就是一个好例子。基地位于层层升起的山脊地区且有树木等因素限制，建筑师利用若干轴线的拓扑性组合布局、引导组群空间，"在平面上，建筑群在透视中相撞，增添了运动感的深度"。这是一座与现代派大学很不相同的校园，她仔细地与当地纹理组合在一起，"平面弯来弯去以避开现有的红杉树"（图10）。建筑师还充分考虑了人在行走过程中的视觉感受，使人感受到自身的重要性，一如舞台上的舞者。詹姆斯·斯特林（James Stirling）的杜塞尔多

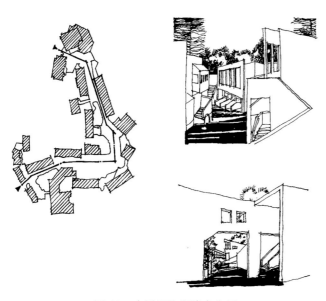

图 10　克累斯格学院宿舍区

博物馆的入口部分也加入了拓扑性转换的因素来增加序列的层次感。

3. 轴线的几何性转换

人的根本力量在于他有知觉和推理能力，因而对天然因素就能做更深入的利用和改造。轴线的几何性转换就是这种力量和表现之一，这种转换的意义是：道德可控制人的行为和心理反应——偏折能激发行人产生与前段轴线所造成的心理期待感有所不同的神秘感，其次通过转折使人的视线产生方向上的突变来组织视觉景观，再次可满足形式构成自身美学上的需要——追求一种更自由的构思新天地，以突破传统静态的永恒美，以期在"多元、混杂、片段、夸张、变形等形式中，追求新的审美肌理"。同时也是为了更灵活地介入到原有的文脉中。

东南大学浦口分校的规划设计，基地大部分是丘陵，作者因势利导，在平原地带和丘陵地带分别使用了几何原则、拓扑原则，入口轴线在中心广场处转折 45°引向丘陵的制高点。从而以富有层次的引导手法将城市空间、过渡空间、中心空间和自然景观组织在一起。南京大屠杀纪念馆的入口序列处理也较好地说明了问题。

应当指出的是：拓扑性转换和几何性转换通常在实践中也都是混合使用的。

三　我们的实践

这是一个设计投标，基地位于三亚市大东海最南端最后一块空地上，用地形状呈南北向狭长走向，总长 500 多 m，形状呈不规则状，业主要求在这块 44062m² 的用地内分别在北区平缓的坡地上布置有 360 间客房的五星级酒店，在南区山地布置单元式别墅，总建筑面积要求达到 7 万 m²。

通过对基地分析，我们认为本设计的难点在于：

1. 如何使这一庞大群体在如此珍贵、显赫的位置上，从形态和空间两方面更有机地介入到现有环境的脉络中，使建筑物与自然物态间富有层次感，在保证大东海区域景观连续性的同时，又体现出该区域的生气和活力。

2. 从基地本身来看，如何使山地别墅区与坡地五星级酒店这两个不同形态的群体在同一基地上取得内在的、有机的关联，从而充分利用本工程狭长的沿海面取得独特的物业形象。

3. 在形式塑造上塑造出层次，以区别于大东海已建的酒店。

看来，两个层次的关联和形式层次感的塑造成了本设计的要旨所在，由于轴线具有联系并组织秩序及丰富形式层次等多种功能，所以我们通过它较有效地解决了以上难点，设计是这样展开：

（1）定位及模式生成：考虑周围各主要视点的视觉要求、内部功能特点及外部空间的特色等因素而引出若干轴线，从而生成总体模式。

（2）内部结构确定：在生成模式的基础上进行内部空间序列及功能关系的处理。考虑到建筑所处环境，我们先确定了一条联系轴（联系南北区）和一条垂直于联系轴的中心轴（统领整个建筑群的控制性轴线），然后针对酒店和别墅群特点，结合山地等高线、坡地走向、海景房视线、总体形象等将中心轴两侧的结构进行脱开、分离、转换、调适（图11、图12）。

（3）形式及空间处理：把平面模式及结构中所产生的想法立体化，并赋予功能上的意义，作为介入到大海与青山之间的建筑群，高架的绿廊、悬索桥、依山就势的别墅和形态

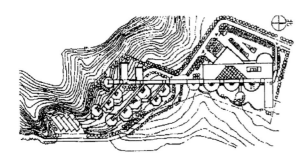

图 11 三亚迎宾馆方案总平面图

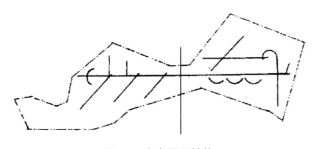

图 12 方案平面结构

别致的客房簇,为大东海的城市景观带来活力,并赋予层次丰富的形象。

这个设计使我们体会到:轴线方法是一个具体而有效的方法,尤其在塑造空间和形式的层次方面,其作用往往会给构思带来新意。

四 结论

针对建筑创作的主体——建筑师来讲,"手法"一词有"主动的、有意识的、系统化的"意思。故本文通过挖掘轴线手法的创造性本质,然后通过具体探讨轴线手法在当代建筑设计中的应用价值和途径,特别是对建筑形式和空间层次塑造方面指导作用的系统分析和比较,目的是为建筑师在作品中表现创意和催化个人的想象力提供一种策略。

本文的探讨一方面是为了消除一些人对轴线这一经典手法的偏见和隔阂,同时也为了使之在当代建筑中发挥潜能。总之,当代建筑文化正处在宏观意义上的扬弃–拓展–重新建构的重要阶段,基于本文对其实际创作中应用的探索,可以预言,轴线手法将会被广大建筑所系统地掌握,并加以不断发展,从而成为一种为建筑师体现创意的有效办法。

参考文献:

[1] 阿恩海姆. 兰术与视知觉. 滕守尧译, 北京: 中国社会科学出版社, 1985.

[2] 彭一刚. 中国古典园林分析. 北京: 中国建筑工业出版社, 1986.

[3] 查尔斯·摩尔. 人体记忆与建筑. 台北: 尚林出版社, 1981.

[4] 罗杰·克拉克. 建筑典例—构形意念之分析与运用. 许丽淑等译, 台北: 尚林出版社, 1984.

[5] 哈特曼. 语言与言学词典, 黄长著译. 上海: 上海辞书出版社, 1984.

本文引改自《工程建设与设计》1998 年第 5 期

Axial approach and hierarchical shaping of architectural space and form

Wang Zhenjun

Form creation is not a mysterious ability, and there are many ways to help people to develop it. In my opinion, among the various ways to create forms, the axial approach is one of the ways with considerable potential.

As we all know, in recent years, on an international scale, architectural classicism has been developed mainly in two aspects: one is the decorative details, and the other is the law of composition. While postmodernism in classicism attempts to resurrect the decorative aspects of classicism, I believe that it is those compositional laws that have the most fundamental and significant influence on the entire architectural landscape of mankind. At the same time, the study of the axial approach, which is one of the classical compositional rules of traditional Chinese architecture, has the practical significance of "maintaining the continuity of architectural culture", which can be figuratively compared to the wiring work.

The establishment of spatial order and hierarchy is the instinctive demand of human spatial experience, which is of great significance to the use and identification of space. "In excellent architecture, the structural, functional, and aesthetic sequences must be organic and compact, and the primary task of architecture is to design with this method of organizing procedures." Traditional Chinese architecture attaches great importance to the shaping of hierarchy, often through the wall, and screen to organize the indoor and outdoor spatial sequences, at the same time the grouping focuses on the pursuit of hierarchical interest through the guidance and transformation of the primary and secondary axes. This paper intends to explore the shaping of form and spatial hierarchy by defining the conversion of axis-like elements and the conversion of axes.

I. Transformation and hierarchical shaping of the elements defining the axis

Transformation can be divided into vertical transformation along the direction of the axis and horizontal transformation relative to the axis.

1. Vertical transformation - detachment and separation

Before we analyze the specific methods, it is necessary to first analyze the visual and psychological basis of the sense of hierarchy. Rudolf Arnheim's study of graphic hierarchy is based on the analysis of the transformational relationship between "figure and base". He pointed out that the formal composition should not only pay attention to the conditions and limitations of the figure (form) separated from the bottom (background), but also should pay attention to the important role of the bottom itself in the formal composition. The formation of the hierarchy is the result of the interaction of the base of the diagram, so the formation of the hierarchy of the base of the diagram, i.e., the longitudinal axis defines the degree of interrelationships between the elements and the degree of linkage, which has become a vertical sequence of factors that can not be ignored. Disengagement and separation in the hierarchy may be a good approach.

The application of this method to the actual creation often stimulates the creativity of architects. For example, in the design of Hefei Shushan Martyrs Monument, the architect used the axis method in the overall layout, and the monument body of the elements defining the axis of the longitudinal disengagement and separation of the treatment, which not only breaks through the conventional, but also well shaped the sequence of the monumental space. The transformation of the traditional "narrative poem" plan layout of monumental architecture, this unconventional treatment made it win the competition. M. Graves' Hanselman House (1967) also utilizes the disengagement of the axial elements, which enriches the hierarchy of the house and constitutes a refreshing image of the house. "The house is layered like a movie, and one can play between the two layers and enjoy the sweetness of family that fills one's heart when one returns."

From the analysis of the above examples, it can be seen that dealing with the relationship between the elements of the figure base has become the key to the shaping of the sense of hierarchy, and the success of the above works lies in:

(1) The use of the role of the axis of orientation and organizational order, strengthening the ephemeral relationship between the elements, so as to convert on this basis, otherwise it is easy to cause the disconnection between the formal level and the spatial level.

(2) The vertical disengagement of the defining elements increases the hierarchy in both spatial and temporal dimensions, thus enriching the meaning of the axes in terms of function. For example, the axes between monuments can prompt the viewer's association and reverie, and the bridge in the residence becomes a channel connecting the main part and the studio.

(3) Formal and spatial echoes between elements. For example, in the Hanselman House, the imaginary corner of the main body suggests the entrance part of the house.

2. Transverse transformation-disconnection

Lateral transformations are mainly formal and spatial transformations of defining elements in the direction perpendicular to the axis. Just as differences in the position of words under the same grammatical structure in linguistics can cause changes in the meaning of a sentence, lateral variation of defining elements on the axis produces another effect.

Lateral disengagement is also called fracture. The Japanese architect Takamitsu Azuma called it "slit". In short, it can be seen as a method of dichotomy formed after splitting a whole. The axial space formed after the disconnection acts as a "harmonizing factor", and its spatial characteristics are not pure, but are formed through juxtaposition, penetration, and other methods of connection. Just as the introduction of simultaneous sounds enriches the structure of music indefinitely, the axial space makes it possible for different elements to be expressed simultaneously, rather than having the characteristics of the two sides disappear at the same time, as in the case of the symmetry of classical compositions due to the connection of the axial space.

Functionally speaking, the horizontal disconnection treatment further improves the overall spatial order in the interior, which is conducive to functional zoning, and at the same time makes the interior and exterior naturally continuous. In the 1982 Paris New Opera House International Design Competition, Hong Kong architect Yan Xunqi used the "disconnection" treatment. From the Bastille Square in the July columns lead to an axis to break the base into two parts: the right is the main part of the symmetrical form in order to maintain the characteristics of the traditional city boundaries; the left is the auxiliary part of the free form. The designers utilized the gap as a commercial street. "This new street is completely pedestrianized and thus allows for a wide range of events ...... giving the famous human touch to the streets of Paris, which seeps into the whole environment and in turn encourages greater public participation in the various performances of the theatre".

Additionally, further treatment at the disconnection point can create a richer sense of spatial hierarchy and facade effect. For example, boldly exposing the contradictions of various aspects formed by structure, railings, and circulation cores is an effective approach. Takamitsu Azuma exposed the distorted second-floor balustrades in the fissure space in his design of the "Wilderness House" residence, enhancing the dynamic and hierarchical sense of the slit . Frank Gehry introduced a fissure on the axis in a courthouse teaching building and extended a staircase and twisted it at the fissure to emphasize the entrance, enriching the facade effect.

3. Simultaneous conversion in vertical and horizontal directions

Above we have explored the transformation from defining the axis elements horizontally and vertically respectively, while in actual creation this transformation is often simultaneous.

Japanese architect Katsuhiro Kobayshi has investigated this approach. Figure 6 shows the compositional analysis he made of his work.

The first step is to define the axes. The whole group is analyzed and divided into geometric volumes, and the axes are defined taking into account the surrounding context.

In the second step, the parts are regularized according to classical compositional techniques such as centripetalism and symmetry.

In the third step, the original elements defining the axes are disengaged and separated in both vertical and horizontal directions, and the elements are further processed into free geometric forms to give a sense of deviation.

In the fourth step, the elements reflected in the facade are further delineated according to the nature of each programme.

The introduction of heterogeneous elements (e.g., fragmentary disorder and difference) into a highly ordered structure is an effective means of changing the static and apathetic uniformity of the work, and of reflecting the individuality and timelessness of the times. The Roosevelt Memorial in Washington, D.C., is an example of the successful application of this method. The architect cleverly combines the monumentality, sense of direction and order of the axis with the indeterminacy and sense of drift of the curved wall, and the curved wall inscribed with text is twisted into different angles by the axis, providing visitors with a multi-angle field of view and a richer amount of information.

4. Relative rotation of the defining element and the axis

By keeping the defining element and the axis in their own state, and then introducing the axis into the defining element at a non-rectangular angle, the work can also be shaped into a rich sense of hierarchy and recognizability. At the same time, this kind of non-right-angled intervention is more likely to bring new ideas and vitality to the composition of the form than right-angled intervention. The relative rotation should be based on the analysis of the original structure.

In Princeton University, there was originally a ramp connecting the train station and gymnasium in the northeast corner. And I.M. Pei, in the design of the student dormitory, based on the convenience characteristics of student behavior and topographical features, determined a common walking path at a 45-degree angle to the entire campus. This maintains the original behavioral and psychological conventions of the students on the one hand, and brings vitality to the design of the entire dormitory complex on the other.

II. Transformation and Hierarchy of the axis

The reason why cities, clusters and single buildings can present various forms and levels is that, in addition to the various expressions of the defining elements themselves, the axes themselves, which organize the defining elements, can also be transformed and combined in a certain way to show their diversity, because their transformation leads to fundamental changes in spatial directions and rhythms.

As far as the transformation of the axis itself is concerned, there are no more than two combinations-geometrical and topological relationships. This linear combination mainly refers to the combination of several axes for the purpose of spatial hierarchy shaping and other factors such as topographical constraints and functional requirements. Generally speaking, a single axis as a linear medium has been throughout, the human mood is only a kind of progression and the amount of accumulation, and through the axis itself can break the monotony of a single axis straight, with a twist to activate the human eye, increase the depth of field in pursuit of qualitative changes. The above method has been applied in the design of ancient Chinese temples and private gardens.

1. Inspiration of ancient mountain forests, temples and private gardens

Subject to the constraints of social and cultural factors, traditional Chinese architectural groups have formed a pattern of consistent central axis and left-right symmetry in general. However, the ancient craftsmen in the long-term practice, through the comprehension of the characteristics of the axis, were able to liberate themselves from rigid and unchanging patterns. They employed flexible techniques such as deflection, displacement, and detour in the use of the axis, adapting to different climates, terrains, and types of conditions. This not only ensured that the construction projects were feasible and economically convenient but also upheld social and cultural conventions. This phenomenon is more prevalent in ancient temples in mountain forests and private gardens.

Figure 9 shows the entrance treatment of a temple in Wutai Mountain, Shanxi. Due to the influence of the terrain and Feng Shui principles, it was not possible to directly face the bustling city with a gate. The craftsmen therefore made a turning treatment to the entrance sequence. Sometimes in order to adapt to the mountain, craftsmen often through a few sections of the axis to organize the group, the axis of the terrain along the winding, in shaping the solemn atmosphere at the same time, but also rendered the temple beyond the world of romance. The well-known Suzhou gardens have fully utilized these techniques.

2. Topological transformations of the axis

Topological relationship refers to a relationship without strict mathematical and geometrical elements, and specifically to the axes, it refers to a linear combination of relationships without a definite angular relationship. Although the topological relationship originates from a simple experience of the environment, which is a kind of low-level, intuitive instinctive treatment of "disorder", in contemporary times, architects have consciously applied it to their creations, which often brings new meanings to the formal composition.

Charles Moor's design for the dormitory area of Kresge College is a good example. Located on a rising ridge and constrained by trees and other factors, the author utilizes the topological combination of several axes to lay out and direct the grouping of spaces, "in plan, the buildings collide in perspective, adding depth to the sense of movement". This is a campus very different from a modernist university, carefully assembled with the local environmental textures, curving around in plan to avoid the existing redwood trees . The designers also took into account the visual perception of people as they walk, making them feel important, like dancers on a stage. J. Stirling's entrance to the Dusseldorf Museum also incorporates topological transformations to add a sense of hierarchy to the sequence.

3. Geometric transformation of the axis

The fundamental power of the human being lies in his ability to perceive and reason, and thus to utilize and transform natural factors in greater depth. The geometric transformation of axes is one of the manifestations of this power, and the significance of this transformation lies in the moral control of human behavior and psychological reactions. First of all, the deflection can inspire the pedestrians to produce a sense of mystery different from the psychological expectation caused by the previous section of the axis, secondly, through the transformation of the human sight to produce a sudden change in direction to organize the visual landscape, and again to satisfy the aesthetic needs of the form of the composition itself, so as to pursue a new world of a freer conception. This can break through the traditional static eternal beauty, and pursue a new aesthetic mechanism in the form of "diversity, hybridization, fragmentation, exaggeration, deformation, etc.", and at the same time, it is also to intervene in the original cultural lineage more flexibly.

The planning and design of Pukou Campus of Southeast University, most of the base is hilly, the architects make the best use of the situation, in the plain and hilly areas respectively, using geometric principles of topological principles, the entrance axis in the central plaza at a 45-degree angle to the heights of the hills, so that a layered approach to guide the urban space, transitional space, the central space and the natural landscape organized together. The entrance sequence of the Nanjing Massacre Memorial Hall is also a good illustration of this type of approach.

It should be noted that topological and geometric transformations are often mixed in practice.

III. Our practice

This is a design competition, the base is located on the last vacant land at the southernmost tip of Dadonghai in Sanya City. The shape of the site is irregular with a narrow north-south orientation and a total length of more than 500 meters. The owner requested to build a five-star hotel by arranging 360 rooms on the gently sloping land in the north area and unit villas on the mountainous land in the south area respectively, with a total construction area of 70,000m^2 in this site of 44,062m^2.

By analyzing the base, we think the difficulties of this design are:

1. How to make this huge group in such a precious and prominent location, from the form and space more organic intervention into the existing environmental veins, so that the buildings and natural objects have a sense of hierarchy between. While ensuring the continuity of the landscape of Dadonghai area, it also reflects the vitality and vigor of the area.

2. From the point of view of the base itself, how to make the mountain villa area and slope five-star hotel two different forms of groups in the same base to obtain the inner, organic connection, so as to make full use of the project's long and narrow coastal surface to obtain a unique architectural image.

3. Shape the level in the form to distinguish it from the built hotels in Dadonghai.

In short, the connection of the two levels and the hierarchy of forms shapes the main purpose of this design. Since the axis has the functions of connection, organizing order and enriching formal hierarchy, we use the above techniques to solve the above difficulties more effectively, and the design is developed in this way:

(1) Positioning and pattern generation: Considering the visual requirements of the surrounding major viewpoints, the internal functional characteristics and the characteristics of the external space, etc., a number of axes are induced to generate a general pattern.

(2) Determination of internal structure: Based on the pattern generation, the internal spatial sequence and functional relationship are handled. Considering the environment in which the building is located, we first determined a contact axis (linking the north and south) and a central axis perpendicular to the contact axis (the controlling axis leading the whole complex), and then for the characteristics of the hotel and villa group, combined with the contour line of the mountain, the direction of the slopes, the view of the sea view room, and the overall image of the central axis of the two sides of the structure of the detachment, separation, conversion, and adaptation.

(3) Formal and spatial treatment: the ideas generated in the planar pattern and structure are three-dimensionalized

and given a functional meaning. As a cluster of buildings intervening between the sea and the green hills, elevated green corridors, suspension bridges, villas on the hillside, and clusters of chic guest rooms bring vitality to the cityscape of the Dadonghai and give it a layered and rich image (Programme Director: Chief Engineer Huang Xingyuan).

This design makes us realize that the axial approach is a concrete and effective method, especially in shaping spatial and formal hierarchies, which often brings new ideas to the conception.

IV. Conclusion

For architects, who are the main body of architectural creation, the word "method" means "active, conscious and systematic". Therefore, this paper explores the creative essence of the axis technique and provides a systematic analysis and comparison of its application value and approaches in contemporary architectural design, particularly in guiding the shaping of architectural form and spatial hierarchy. The aim is to provide architects with a strategy to express their creativity and catalyze their imagination in their works.

The purpose of this paper is, on the one hand, to remove the prejudice and disconnection that some people have towards the classic technique of axis, and, on the other hand, to enable it to realize its potential in contemporary architecture. In conclusion, contemporary architectural culture is in an important stage of abandonment, expansion and reconstruction in a macro sense. Based on this paper's exploration of its application in actual creation, it can be predicted that the axial approach will be systematically grasped by the architectural industry and continuously developed, thus becoming an effective way for architects to manifest creativity.

References:
[1] Arnheim. Orchidology and Visual Perception. Translated by Teng Shouyao, Beijing: China Social Science Press, 1985.
[2] Peng Yigang. The analysis of Chinese classical gardens. Beijing: China Architecture Industry Press, 1986.
[3] Charles Moore - Body, Memory, and Architecture. Taipei: Shanglin Publishing House, 1981.
[4] Roger Clark. Analysis and Application of Conformational Ideas in Architecture - Hsu, Li-Shu, et al. Taipei: Shang-Lin Publishing House, 1984.
[5] Hartmann - Dictionary of Language and Linguistics, translated by Huang Changzhi. Shanghai: Shanghai Rhetoric Publishing House, 1984.

This article is adapted from *Construction & Design for Engineering*, Issue No. 5, 1998.

建筑的本质与中国建筑师的文化使命

王振军

当下中国建筑界呈现出两种现象，一方面我国以完全开放的姿态引入了大量优秀的欧美建筑事务所和国际建筑师来参与并完成了大量的重要项目，中国已经完全融入了国际化的建筑语境中，2008奥运会和2010年上海世博会就是最好的例证，另一方面现代主义建筑的传统和基本内核尚未在中国完全、真正地确立，同时中国建筑文化的发展和繁荣也还有大量的工作要做。针对上述两种现象，结合今天会议主题，我谈如下几点体会：

（1）中国丰厚的建筑文化积淀和改革开放30年大量的建筑设计实践，应让我们更自信地去展示中国建筑师的风采，勇于参与国际化的竞争。北京奥运会和上海世博会中国建筑师的表现已经初步展示并证明了我们的潜力。

（2）建筑设计是对当下社会生活和社会变革的一种反映，建筑创新就是基于当下大众生活需求的一种发现。建筑创作并不仅仅是几何形式的智力游戏，而是要从当下人们的生活需求和社会发展入手，去创作符合他们当下生活方式的建筑作品。比如说现代展览建筑，由于多媒体技术和显示技术对人们生活的渗透，人们已经不满足于传统的方式而要求在参观中能体验到更逼真、更震撼、更有感染力的感知方式，人们与展示空间需要更真诚的互动与共鸣。传统的设计语言似乎已不能够满足高科技时代人们观展时的精神和情感需求。多年来的建筑实践使我们更坚定了原来的设计观念：从人们生活需求的本质出发，同时关照文化传承，让建筑多一些融合、少一些炫耀。

（3）建筑是文化的容器，承载着文化传承和推广的使命，责任重大，但也需要全社会的共同努力。建筑，无论是其形式还是空间都是文化要素的载体和容器，因此建筑比起其他艺术品更具社会和文化属性，因此好的建筑无一不是建筑师在充分尊重、理解和欣赏并且融入业主的文化需求的前提下得出的结果。同时，建筑作为艺术创作的形式之一，由于其强烈的社会属性使其产生的过程必定是超出了建筑师个人的审美范畴，好作品的产生也在很大程度上依赖于社会大众的感知、评判和接受程度以及审美素养。好的作品的产生一定是在以建筑师为主导下，建设各方通力协作的结果。

（4）中国建筑的年轻建筑师需要引导，"到底什么是好建筑？""建筑的本质是什么？"缺乏引导会导致同质化和西方化。目前一些建筑作品中确实呈现出"以他人之新为新""鹦鹉学舌"的趋势和现象。技巧和手法固然重要，但如果我们自我思考和理念缺失，长此以往会使建筑创作永远停留在技巧和手法的层面，而最终迷失在西方的评判标准中，将西方升华为偶像的过程就是自我弱化的过程。由此我们需要系统性和规范化地总结中国建筑30年来创作实践，从而才能从理念到实践上指导建筑师的下一步的创作活动。

本文引改自作者在2012中国建筑学会"繁荣和发展中国建筑文化座谈会"发言和《建筑学报》2012年第2期文章

The Essence of Architecture and the Cultural Mission of Chinese Architects

Wang Zhenjun

In the contemporary Chinese architectural field, two notable phenomena can be observed. On one hand, China has embraced an open stance, inviting numerous excellent European and American architectural firms and international architects to participate in and complete a significant number of important projects. This has effectively integrated China into the global architectural discourse, with the 2008 Beijing Olympics and the 2010 Shanghai World Expo serving as prime examples. On the other hand, the tradition and fundamental core of modernist architecture have not yet been fully and authentically establ China. Furthermore, there remains substantial work to be done in the development and flouri shing of Chinese architectural culture. In light of these phenomena and the theme of today's conference, I mould like to share the following observations:

1.The rich architectural cultural heritage of China and the extensive practical experience in architectural design practice accrued over 30 years of reform and opening-up should inspire greater confidence among Chinese designers to showcase their talent and courageously engage in international competition. In the Beijing Olympic Games and the Shanghai World Expo, the performance of Chinese architects has shown and proved our potential.

2.Architectural design serves as a reflection of contemporary societal life and its transformations. Architectural innovation is essentially a discovery based on the current demands of the public's lifestyle. Architecture creation is not merely an intellectual game of geometric forms but rather an endeavor that starts from the current needs of people's lives and societal development, aiming to create architectural works that align with their current way of life. For instance, in modern exhibition architecture, due to the infiltration of multimedia and display technologies into daily life, people are no longer content with traditional methods and instead demand to experience more realistic, impactful, and engaging modes of perception during visits. There is a need for genuine interaction and resonance between individuals and exhibition spaces.

Traditional design languages may no longer fulfill the spiritual and emotional needs of people in the high-tech era.

Years of architectural practice have reinforced our original design concept: to start from the essence of people's needs in daily life, to care for cultural heritage, and to create architecture that integrates more and boasts less.

3.Architecture serves as a vessel for culture, bearing the mission of cultural inheritance and promotion, thus carrying significant responsibilities.However, this endeavor also necessitates collective efforts from society as a whole.

Architecture, both in its form and spatial configuration, serves as the vessel and container for cultural elements. Consequently, architecture possesses a greater societal and cultural attribute compared to other forms of art.Thus, exceptional architecture invariably results from architects fully respecting, understanding, appreciating, and integrating into the cultural needs of clients.

Moreover, as a form of artistic creation, architecture, due to its strong social characteristics, undergoes a process that transcends the personal aesthetic boundaries of architects. The creation of exemplary works largely depends on the public's perception, judgment, acceptance, and aesthetic literacy. Exceptional works are invariably the result of collaborative efforts led by architects among all stakeholders.

4. Young Chinese architects require guidance on defining "what constitutes good architecture" and "the essence of architecture". A lack of guidance leads to homogenization and Westernization. Currently, certain architectural works indeed exhibit trends and phenomena characterized by the tendency to adopt "novelty for the sake of novelty" and "parroting of others ideas". Skills and techniques are undoubtedly crucial, but if self-thinking and ideas are missing, in the long run, a perpetual focus on them risks

confining architectural creation solely within the realm of technical proficiency. Consequently, there's a risk of ultimately losing oneself within Western standards of judgment, whereby the process of idolizing the West becomes a process of self-weakening. Therefore, it is necessary to systematically and formally summarize 30 years of Chinese architectural creative practice to guide architects' next steps in creative activities from conceptualization to practice.

This article is adapted from the author's speech at the 2012 China Architecture Society "Symposium on the Prosperity and Development of Chinese Architectural Culture" and from the article in the second issue of *Architectural Journal* in 2012.

质疑的时代更需要标准

王振军

在充满激情、幻想、追求的21世纪，世界的大问题其实也就是设计界的大问题，而等待回答的问题又比以往任何历史时代都多、都乱、都难。

前段时间，中国建筑学会邀约参加"什么是好建筑"的讨论，我欣然答应了，因为，自己虽然已经从事建筑设计有近30年的时间，但从心底里也还是需要再次确认一下，在当下"究竟什么样的建筑才能称得上是好建筑"。

在这样一个多元的、充满质疑的年代来讨论标准问题肯定不是一件轻松的事情，但毫无疑问，高速生成的新建筑和不断冒出来的社会大众对相当一部分建筑提出的批评和戏说、诟病这一现象，的确使我们发现这一现象其中大多是由于建筑越来越不像建筑，或者说越来越偏离了建筑的本质而导致的。由此还涉及对建筑的感知、创新的理解以及建筑的社会属性、内容与形式的关系等相关话题。

因此顺着"建筑本质"这个思路来讨论"什么是好建筑"似乎是符合理性、思辨的好方法。我们首先可以说：反映了建筑本质的建筑才是好建筑。

什么是建筑的本质

老子在两千多年前讲到"有无相生"理论的时候，已经认识到做建筑其实是在做空间。我在做沙特馆的时候，觉得我们的先贤太伟大了，空间就是主角，而在此应该强调的是人又是空间的主角，一个空间再宏大，实际上是靠人来体验的。这就是挪威哲学家克里斯蒂安·诺伯格-舒尔茨（Christian Norberg-Schulz）所提出的"场所精神"，场所是要人来体验的，人和空间组合在一起才能产生真正的建筑体验。虽然现在有人说，当代建筑已进入一个自由时代或者说试验的时代，但不管怎样，空间是建筑的主角，人又是空间的主角，这样一个命题应该是永远不变的。

建筑感知度

"感知度"是信息论中的一个名词，是指人对信息的接近程度。2010年上海世界博览会沙特馆为什么会被观众追捧，说明它的感知度比较高。从《一千零一夜》故事提出的元素，包括海上丝绸之路，大家都非常熟悉，国外建筑师他们小时候也看过《一千零一夜》。一个建筑的构思应该具有比较高的感知度。在建筑层面当中，寻找一个高感知度的元素更能引起共鸣，而建筑故事不要讲得太晦涩，特别是公共建筑。

对建筑创新的理解

设计是对社会变革的一种反映，这句话说得非常深刻。沙特馆的参观方式，老百姓比较喜欢，正是因为我们这种展示的方式呼应了高科技时代观众对参观博物馆的一种需求。它颠覆了传统二元并置式的观看模式，创造了一种更逼真、更融入或者说更震撼的参观体验。从这点上来说，创作时不应把太多的精力花在外形的时尚上，时尚的东西是容易过时的，本质的东西才是永恒的。

建筑的社会属性

建筑师不应把建筑当成塑造其个人特征的手段。由于其强烈的社会属性，故不能用评价其他艺术形式如雕塑、绘画的方法来评价建筑，它还是一种空间的艺术，它有很多文化的使命。建筑已经脱离了纯个人的美学范畴，因为对建筑的价值判断已经融入了业主、包括大众等各个层面人士的意见和判断。

形式与内容的关系

沙特馆如果没有船的外形，也不可能有立体、全景融入式的流线设计。因此内容和形式谁决定谁已不重要，重要的是内容与形式的统一，并能高度融合进而产生一种在两者之上的新的设计语言。

好建筑应该是真实的建筑

什么是好的建筑？总结起来，我想第一是在设计当中投入了真实的感情。第二是概念比较清晰，能引起大家共鸣。第三是形式和内容高度统一。第四是空间体验生动感人。第五是要关照文化传统，因为建筑承载着很多文化的作用，要体现文化建筑，这是建筑应担负的责任。

本文引改自《建筑与文化》2013年第2期

The Necessity of Standard in Oppugning Time

Wang Zhenjun

In the passionate, fantastical, and aspirational 21st century, the world's big questions are, in fact, also the major issues confronting the design industry, and the questions to be answered are more numerous, chaotic, and difficult than in any historical era.

Recently, the Architectural Society of China invited me to participate in a discussion on "What is Good Architecture?" I gladly accepted because, even after nearly 30 years of architectural design practice, I still felt the need to re-confirm, deep in my heart, what actually constitutes good architecture in this current era.

To discuss the issue of standards in such a diverse and questioning age is certainly not an easy task. Undoubtedly, with the emergence of rapidly constructed new buildings and the criticisms, jests, and grievances voiced by the general public towards certain architectural works, we realize that this phenomenon is mostly due to the fact that it is becoming less like architecture or deviating from its essence. This occurrence has sparked discussions on a range of related topics such as architectural perception, innovation comprehension, the social attributes of architecture, and the relationship between content and form.

Thus, addressing "what is good architecture" from the perspective of the "essence of architecture" seems like a rational and thoughtful approach. We can first say that only architecture that reflects its essential nature can be considered good architecture.

What Constitutes the Essence of Architecture

Over two millennia ago, Laozi articulated the theory of "mutual arising of existence and non-existence", recognizing that the act of architecture is essentially the creation of space.

When I was working on the Saudi Pavilion, I couldn't help but marvel at the greatness of our predecessors. The space takes on the leading role, yet it is crucial to emphasize that humans are also protagonists within this space. Regardless of how vast a space may be, its true essence lies in the human experience it facilitates. This aligns with what the Norwegian philosopher Nils Schultz termed the "spirit of place"—a place demands to be experienced by people, and it is the combination of people and space that creates a true architectural experience. Although some argue that contemporary architecture has entered an era of freedom or experimentation, regardless of the context, the premise that space is the protagonist of architecture and humans are the protagonists of space remains an unchanging truth.

Perceptibility in Architecture

"Perceptibility" is a term from information theory that denotes the degree of accessibility of information to an individual. The high popularity of the Saudi Pavilion at the World Expo underscores its significant perceptibility. The elements drawn from the "One Thousand and One Nights", including the maritime Silk Road, are very familiar to many, even among foreign architects who recall these stories from their childhood. The concept of a building should possess a high degree of perceptibility. Within the realm of architecture, identifying elements with high perceptibility can foster resonance. The story of architecture should not be too obscure, especially in public buildings.

Understanding of Architectural Innovation

Design is a reflection of societal change—a profound statement. The popularity of the exhibition format in the Saudi Pavilion suggests that it resonates with the audience's expectations in a high-tech era. This innovative approach to exhibition subverts traditional binary oppositional viewing patterns, creating a more immersive, integrated, or even more shocking visiting experience. From this perspective, when designing, one should not invest too much energy in the superficial aspects of fashion, as fashionable things tend to become outdated, while essence is what is eternal.

The Social Attributes of Architecture

Architects should not regard architecture as a means to express their personal characteristics. Given its strong social attribute, architecture cannot be evaluated using the same criteria applied to other art forms like sculpture or painting. Architecture is a spatial art with many cultural responsibilities. It has moved beyond purely personal aesthetics, with value judgments about architecture now influenced by a range of stakeholders, including clients and the general public.

The Relationship Between Form and Content

If the shape of a ship like the Saudi Pavilion lacked, it would be impossible to achieve 3D panoramic integrated circulation design. Therefore, it's no longer crucial to determine which between content and form is decisive; what matters is the unity of content and form, and their ability to blend harmoniously, creating a new design language that transcends both.

Good Architecture Should Be Authentic

What constitutes good architecture? In summary, the first factor is genuine emotion invested in the design process. The second is a clear concept that resonates with a wide audience. The third is a high degree of coherence between form and content. The fourth is a vivid and engaging spatial experience. Finally, a good architecture design should consider cultural traditions, as architecture carries significant cultural responsibilities. It is the responsibility of architects to reflect cultural heritage in their designs.

This article is adapted from *Architecture & Culture*, Issue No. 2, 2013.

像做博物馆一样做工业建筑

王振军　郑秉东　李达

引言

工业建筑与博物馆建筑虽然在功能和尺度上有一定的差异，但他们都同样具有一定的建筑体量、大面积的实墙立面，高大的室内空间，以及由此构成的体量感和标志性。近些年来，由于这两种类型的建筑我们都同时在做，特别是工业项目的甲方对其建筑的美学要求越来越高，使我们意识到"像做博物馆一样做工业建筑"越来越有必要，而且工业建筑也完全可以通过努力实现，塑造出像博物馆一样具有高度美学价值、生动感人的建筑作品。

1 博物馆与工业建筑的异同点

1.1 博物馆与工业建筑的相同点

（1）立面实墙多，开窗少。博物馆与工业建筑（特别是有洁净要求的工业建筑）受内部功能影响，立面开窗较少，可以实现大面积的实墙面，使建筑具有很强的体量感与冲击力，立面造型简洁纯净。

（2）功能流线要求高。博物馆与工业建筑不同于一般的民用建筑，其内部功能流线的组织排布都有较高的要求。博物馆的核心是展览流线，工业建筑的核心是工艺生产流线。人与机器的动线设计是博物馆与工业建筑规划布局的关键因素。

（3）体量上高度方面有限制。博物馆因为疏散问题，工业建筑因为运输问题，高度一般不会太高，但在水平方向都有较大的建筑体量和较长的建筑界面。特别是工业建筑其巨大的体量在城市尺度下具有很强的视觉冲击力。

（4）主要空间尺度相似。博物馆需要满足高大空间的展览需求，工业建筑需要满足大型设备的生产需求，他们都具有高大的室内空间。

（5）艺术感。博物馆建筑具有很高的艺术美感，例如蓬皮杜艺术中心高技派的设计手法就与工业建筑十分相似。而高水准的工业建筑也有很强的工业美感，第一座现代主义建筑——德国通用电气的透平机车间就是明证。

（6）标志性。博物馆建筑与工业建筑以其巨大的体量和视觉冲击力，往往很自然地成为一个区域的地标建筑。

1.2 博物馆与工业建筑的不同点

（1）功能要求不同。博物馆是文化类的民用建筑，优先考虑的是人的观展体验。工业建筑优先要满足工业生产的需求，在内部功能的排布上有显著差异。

（2）造型自由度。博物馆的造型设计自由度更大，观展流线相对灵活，可以形成丰富的造型。工业建筑受生产工艺要求的限制，建筑形体设计的自由度相对较小，但会有一些独特的工业元素，如烟囱，吊装口，管架，大型设备等。

（3）造价不同。博物馆造价相对较高，可以应用更多石材、幕墙、金属板等造价高的立面材料。工业建筑由于土建费用在总造价中所占比重较小，所以平摊到的土建单方造价也就非常少。

（4）结构参数要求不同。博物馆建筑的结构荷载相对较小，能满足一般展览需要即可。工业建筑结构荷载要求较大，需要满足大型设备的荷载要求，一些特殊工艺的工业建筑还需要考虑防微震处理。

台积电南京晶圆厂就充分体现了工业建筑的艺术美感。建筑设计以圆形的几何形体，象征台积电的高阶晶圆生产技术。建筑以极具未来感的几何形式与银白金属外壳，飘浮于森林之间，宛如一座绿林中的晶圆博物馆。

贝聿铭设计的美国国家美术馆东馆，是博物馆建筑的典型代表，体现出极高的艺术性。建筑立面开窗很少，大面积的实墙面形成丰富的几何体量，立面造型简洁纯净，具有强烈的视觉冲击力和艺术美感。内部空间按照观展流线有序排布互相交融，形成流动的空间动线。合理的空间功能性与独特的建筑艺术性完美结合，使东馆成为美国的标志性建筑。

2 设计实践——紫光南京集成电路总部基地项目

我们设计的紫光南京集成电路总部基地项目就是一座像博物馆一样的工业建筑，在工艺及机电设计上达到了国际先进水平，同时在建筑上又体现出很好的艺术效果。该项目总投资700亿元，建设12英寸存储器生产线，设计产能为10万片/月，建设用地面积为39.59公顷，总建筑面积约为54万 m²。主要生产3D NAND Flash储存器芯片，产品达到同类产品的国际先进技术指标。号称"现代工业粮食"的芯片是数字经济的产业核心，该工厂的建成对应对西方在芯片领域的保护主义，发展本土芯片工业具有重要的战略意义。

该建筑设计在尊重芯片生产工艺及动力传输、材料运输等各种需求的前提下，强调现代高科技厂区高效、便捷、智能的同时，更加注重通过丰富的高品质室内外工作空间的打造，来激发员工交流交往的积极性，从而提高研发人员的工作主动性、创造力。

项目依据周边交通、现状地形情况，在合理组织生产流程、满足芯片生产工艺及动力传输、材料运输等各种需求的前提下，将全厂分为厂前支持区、核心生产区和生产辅助区。不同功能的建筑组团遵循生产动线平行排布，形成统一的秩序感。根据相应的功能特性，特别是厂区内员工的活动需求，遵照有机建构的理念来对场地、工艺、空间、材料进行组合，精心处理现代化厂区中人、生产环境、工艺设备及生产效益诸方面的辨证关系。

（1）根据FAB核心厂房剖面，提取设计原型。核心厂房包括设备机房层、技术夹层和洁净生产层，两端为辅助支持区，内部功能自然形成一个U形的剖面结构。厂区建筑的设计提取核心厂房的U形形态作为原形。

（2）通过系统的建筑语言，以U形为基本单元，根据不同厂房的功能形成几个大小体量的U形组合。U形两端高出的女儿墙可以遮挡屋面设备用房。建筑群面向景观河道形成400多m长的建筑立面，高低起伏，生动有序。

（3）不同功能的厂房平行排布，厂房之间通过工艺连廊横向串联，保证了不同厂房之间人流、物流的联通，使工艺流线高效便捷。

（4）根据各栋建筑的功能特性赋予不同通透性的立面材料。运维楼以玻璃幕墙与点窗组合，实现了良好的采光效果和景观视野。厂房立面开窗较少，通过铝板、穿孔板、波纹板等材料的组合搭配，形成统一而丰富的立面效果。工业厂房高耸的实墙与办公建筑通透的玻璃幕墙形成鲜明的对比。

（5）根据生产及其动力需求，以及检测、研发、参观、配套等功能特点，有序组织内部功能，厂前区与生产区分区明确，两者之间由绿化带分隔。将景观元素植入到工业园区中，打造了一座生长在绿林碧波之上的现代工业园区，形成一个有机的、可持续的、具有紫光特色的芯片工厂体系。

厂区总平面布置规整有序，提取核心厂房的U形形态作为母体，来组织具有超长、超大体量和多达26种功能特性的建筑群，并通过具有精细工业美学精神的墙身构造和材料，充分表现高科技工业园区特殊的美学特质。不同功能的厂房高低错动，秩序井然，在统一的韵律中呈现出丰富动人的立面效果。厂前区通过架空和悬挑与厂区景观相融合，以材料的通透性组合来体现其开放性。厂前区环绕办公研发楼设置1.2公里长的带状景观花园，通过水景、竹林、小径和休闲设施打造花园式工厂，充分关注了员工工间休息的需求体验。面向城市主干道，建筑以600m水平延展的超长建筑体量，气势磅礴、宏伟壮观，将工业建筑的恢弘与博物馆的艺术美感结合得淋漓尽致，是一座兼具严整的功能性与极高的艺术性的工业建筑作品。

我们在建筑创作中提出：打破工业和民用界限的"大建筑观"，提出"像做博物馆一样做工业建筑"，尤其是在新时期两者对空间功能性和艺术性的追求是同等重要的。在项目中建立"双总设计师制"，建筑总师全程领衔参与完成了武汉国家存储器基地项目（一期）、紫光集成电路基地、中芯京城集成电路生产线、厦门天马光电子有限公司生产线、深圳证券南方中心、大兴国际机场车辆维修中心，中原云大数据产业城，上海临港芯片厂区等工业项目，在保证生产工艺需求的前提下，通过引入先进的"有机建构"设计理念和"技术、艺术一体化"的设计路线，塑造出了技术与艺术相生共融，具有工业美学特点的现代高科技工厂样板。

结语

以上思考和实践同时也表明：随着信息时代的到来，民用建筑与工业建筑的界限越来越模糊，传统工业建筑的概念在转变，新兴的建筑类型在不断涌现，如数据中心、研发中心、中试平台等。时代的变化需要建筑师树立全新的建筑观,我们就此提出来"大建筑观"。所谓"大建筑观",就是要提倡建筑师树立"超内容、超类型"的大建筑视野,是致广大、尽精微的,是创造为上、走向未知的。"大建筑观"是一种格局和气势,应该是工业院建筑师的选择和价值取向。

遵循建筑学基本要素要求及其外延要求,是所有建筑类型创作中建筑师必须遵循的基本职业准则,不管什么建筑,其本质是一样的,而类型只是具体设计内容不同而已。摒弃惯性的、简单化的二元分类做法,消除对建筑类型的成见,设计院特别是工业院出身的设计院,一定会抓住当前的市场契机,创造出更多的精品工业建筑,迎来更新更大的发展。

本文引改自《中国建筑学会工业建筑第十四届（2023年）年会论文集》

Design Industrial Buildings As Museums

Wang Zhenjun

Abstract: Although industrial and museum architectures differ in function and scale, they share similarities in terms of substantial architectural mass, extensive use of solid facades, and expansive interior spaces, which contribute to a sense of volume and iconic presence.In recent years, as we've been working on both types of buildings simultaneously, we've noticed a growing demand from industrial clients for higher aesthetic standards in their architecture. This has led us to realize the increasing importance of "treating industrial buildings like museums". Industrial structures can indeed be transformed into architectural pieces with high aesthetic value and compelling narratives, akin to museums, through dedicated efforts.

1.Similarities and Differences Between Museum and Industrial Building

1.1 Similarities Between Museum and Industrial Building

1) Facade Design: Both museums and industrial buildings (especially those requiring clean environments) are influenced by their internal functions, leading to fewer windows and more solid wall surfaces. This design approach imparts a strong sense of mass and impact, with facades characterized by simplicity and purity.

2) Functional Flow Requirements:
Unlike typical civilian buildings, museums and industrial buildings have high demands for internal functional flow organization. Museums focus on exhibition flow, while industrial buildings prioritize process production flow. The design of movement paths for people and machinery is crucial in the planning and layout of both types of buildings.

3) Volume and Height Limitations:
Due to evacuation issues in museums and transportation considerations in industrial buildings, heights are generally not excessive. However, both types of buildings exhibit substantial horizontal volumes and extended building interfaces, particularly industrial buildings, whose massive volumes have a significant visual impact at the urban scale.

4) Similar Main Spatial Dimensions:
Museums need to accommodate large exhibition spaces, while industrial buildings need to house large machinery for production purposes; both require expansive interior spaces.

5) Aesthetic Sensibility:
Museum typically embodies a high level of artistic beauty. For instance, the high-tech design approach of the Pompidou Centre closely resembles that of industrial buildings. Similarly, high-standard industrial buildings also display a strong industrial aesthetic, with the first modernist building, the General Electric turbine hall in Germany, serving as a clear example.

6) Iconicity:
Both museum and industrial buildings, with their massive volumes and visual impact, naturally tend to become landmark structures in a region.

1.2 Differences Between Museum and Industrial Building

(1) Functional Requirements:
Museums, as cultural civilian buildings, prioritize visitor exhibition experiences. Industrial buildings, on the other hand, prioritize industrial production needs, resulting in significant differences in internal function layout.

(2) Form Design Freedom:
Museums generally have greater freedom in form design, allowing for flexible and varied exhibition flows and rich architectural forms. Industrial buildings, constrained by production process requirements, have less freedom in building shape design, though they feature unique industrial elements such as chimneys, hoists, pipe racks, and large equipment.

(3) Construction Costs:
Museums typically have higher construction costs, allow-

ing for the use of more expensive facade materials like stone, curtain walls, and metal panels. In contrast, civil construction costs constitute a smaller proportion of total costs for industrial buildings, resulting in tighter per-unit construction budgets.

(4) Structural Requirements:

Museum buildings have relatively lower structural loads, sufficient for typical exhibition needs. Industrial buildings require higher structural loads to support large machinery, and some specialized industrial buildings must also consider anti-microvibration treatments.

TSMC's Nanjing wafer fab exemplifies the artistic beauty of industrial building. Its design features circular geometric variations, symbolizing TSMC's advanced wafer production technology. The building, with its futuristic geometric forms and silvery metallic shell, appears to float among the forests, resembling a wafer museum nestled in the woods.

The East Building of the National Gallery of Art, designed by I.M. Pei, is a quintessential example of museums, reflecting high artistic value. The building's facade features few windows, with large areas of solid walls creating rich geometric volumes. The facade's simple and pure design offers a strong visual impact and aesthetic appeal. Its interior spaces are orderly arranged according to exhibition flows, blending seamlessly to create dynamic spatial dynamics. The rational functionality of spaces and unique architectural artistry combine perfectly, making the East Building a landmark of America.

2. Design Practice: The Tsinghua Unigroup Nanjing Integrated Circuit Base Project

The design of the Tsinghua Unigroup Nanjing Integrated Circuit Base is emblematic of treating industrial building with the sensibilities of museum design. It achieves an international standard in process and electromechanical design while also exhibiting exceptional artistic quality. The project, with a total investment of 70 billion yuan, includes the construction of a 12-inch memory production line with a design capacity of 100,000 wafers per month, occupying a land area of 39.59 hectares and a total construction area of about 540,000 square meters. Primarily producing 3D NAND Flash memory chips, the products meet the advanced international technology standards of similar products. Known as the "modern industrial food", chips are central to the digital economy, and the construction of this factory is strategically significant for developing local chip industries in response to Western protectionism, in the field of chips.

The architectural design respects the requirements of chip production processes and power and material transportation, emphasizing a modern, efficient, and intelligent high-tech campus while fostering high-quality interior and exterior workspaces that stimulate employee interaction and creativity.

The project layout is informed by surrounding traffic and existing terrain. Organizing production processes and meeting various needs such as chip production processes and material transportation, the project is divided the plant into a front support area, core production area, and production auxiliary area. Different functional building clusters follow a parallel arrangement of production flows, creating a sense of unified order. According to the specific functional characteristics, especially the activity needs of employees within the plant, the site, processes, spaces, and materials are combined following an organic construction philosophy, carefully managing the dialectical relationships between people, the production environment, process equipment, and production efficiency in a modern industrial campus.

(1) The design prototype is extracted based on the FAB core plant section, which includes an equipment room layer, a technical mezzanine, and a clean production layer, with support areas at both ends forming a U-shaped sectional structure. The design of the plant buildings extracts this U-shaped form as the prototype.

(2) Using a systematic architectural language, the U-shape serves as the basic unit, forming several U-shaped combinations of varying sizes according to the functions of different plants. The parapet walls at the U-ends conceal rooftop equipment rooms. Facing the landscape river, the buildings create a facade over 400 meters long that is vibrant and orderly.

(3) Different functional plants are arranged parallel to each other, connected laterally by process corridors, ensuring efficient and convenient flow of people and materials between buildings.

(4) Different transparency materials are used for facades based on the function of each building. The maintenance building uses a combination of glass curtain walls and punctuated windows to achieve good lighting and scenic views. Plant facades have fewer windows and use a combination of aluminum panels, perforated plates, and corrugated panels to create a unified yet rich facade effect. The towering solid walls of industrial plants contrast sharply with the transparent glass curtain walls of office buildings.

5) Organizing internal functions according to production

and power needs, as well as testing, R&D, visitor, and support functions, the front area and production area are clearly divided, separated by a green belt. Landscape elements are incorporated into the industrial park, creating a modern industrial park that grows among green woods and blue waves, forming an organic, sustainable, and characteristic Tsinghua Unigroup chip factory system.

The overall layout of the plant is orderly, extracting the U-shaped form of the core plant as the matrix to organize this building group, which has a super-long, massive volume and up to 26 functional characteristics. Through walls constructed with a fine industrial aesthetic spirit and materials, the special aesthetic qualities of a high-tech industrial park are fully expressed. Different functional plants are staggered in height, orderly, and present a rich and moving facade effect. The front area integrates with the landscape of the plant through overhangs and cantilevers, using the transparency of materials to reflect its openness. A 1.2 km long strip garden surrounds the office R&D buildings, incorporating water features, bamboo groves, paths, and recreational facilities to create a garden-style factory, paying close attention to the rest needs of employees.

Facing the main urban thoroughfare, the building stretches horizontally for 600 meters, exhibiting a grand and magnificent presence. Combining the grandeur of industrial architecture with the artistic beauty of museums, it embodies both rigorous functionality and exceptional artistic quality.

In our studio's architectural creation, we propose the concept of the "Macro Vision of Architecture" that breaks the boundaries between industrial and civil buildings, advocating to approach industrial buildings with the same attention to spatial functionality and artistic pursuit as museum design. Particularly in the new era, both aspects are equally important. In our projects, we have established a "dual chief designer system" where the chief architect leads and participates in the entire process of projects, including the Wuhan National Memory Base Phase I, Tsinghua Unigroup Integrated Circuit Base, Zhongxin Jingcheng, Xiamen Tianma, Shenzhen Securities South Center, Daxing International Airport Vehicle Maintenance Center, Zhongyuan Cloud Big Data Industrial Park, and Shanghai Lingang Chip Factory. While meeting the requirements of production processes, we have shaped modern high-tech factory prototypes that integrate technology and art by introducing advanced "organic construction" design concepts and an integrated approach to technology and art. These prototypes embody a harmonious coexistence of technology and art, with distinctive industrial aesthetics.

Conclusion:

The above considerations and practices also indicate that with the advent of the information age, the boundary between civil buildings and industrial buildings is becoming increasingly blurred. The concept of traditional industrial buildings is changing and new types of buildings are emerging, such as data centers and R&D centers. The changes of the times require architects to establish a new architectural perspective, and we propose the concept of the " Macro Vision of Architecture". The so-called "big architecture" advocates architects to establish a "broad and comprehensive" architectural vision, which is inclusive and meticulous, creative and unknown. Big architecture is a pattern and momentum, and it should be the choice and value orientation of industrial architectural designers.

Adhering to the basic requirements of architectural elements and their extension requirements is the basic professional criterion that architects must follow in the creation of all types of buildings. Regardless of the type of building, its essence is the same, and the type is only a difference in specific design content. Abandoning the inertial, simplistic binary classification approach and eliminating biases towards building types, design institutes, especially those with industrial backgrounds, will seize the current market opportunities and create more exquisite industrial buildings, ushering in renewed and greater development.

This article is adapted from *the Proceedings of the 14th Annual Conference on Industrial Architecture by the Architectural Society of China*, 2023.

高科技园区规划设计和建设的思考

——访中国电子工程设计院股份有限公司总建筑师王振军

我国第一座软件园——上海浦东软件园（一二期）是由中国电子院总建筑师王振军主持完成，之后他又带领团队在全国各地完成数十项包括软件园在内的各类高科技园区，2013年由他主编的我国第一部科技园区设计规范——《软件园规划设计规范》颁布发行，同时他将园区设计体验予以总结，撰写十几篇论文发表于《建筑学报》等主流学术期刊。2017年9月《建筑技艺》(AT）杂志社刘小楠、朱晓琳对他进行了专访。

AT： 在国内，大、中型城市中出现了大量的产业园区，您认为应如何定义高科技园区，而它又具备哪些特征？

王振军： 一个园区的规划建设，无论占地10万 m^2 还是100万 m^2，都会对所在的城市周边产生一定的影响——如人流量、交通流量、市政建设、城市空间结构、产业业态等，最重要的是科技园区作为一个城市中新的空间形态，已经对城市结构造成了很大程度的影响，科技园区已与CBD形成城市的两极，因此建筑师有必要更加关注和思考怎样把创新的内容通过高科技园区的规划设计真正带入到城市中。

此外，高科技园区的建设不能只考虑创新，还应当充分思考和理解当地城市、文化脉络，例如在苏州、青岛和哈尔滨分别设计高科技园区，即使前期规划的业态和模式一样，但最终呈现出的建成效果肯定也是不同的。再例如北京市通州区要建设北京市副中心，北京市委、市政府要整体搬迁过去，必然需要新策略。我认为对于园区的建设，比较重要的是探讨和研究人与自然的关系，具体可以概括为处理好以下三个要素之间的关系：人、物理环境、科技与生产方式。1998年，我发表了一篇题为《人与自然通过科技的整合》的文章，文中的核心思想正是这个理念。因此，建筑师的任务就是通过建筑语言和景观语言，把这三个要素整合起来，这是需要解决的最根本问题和出发点。科技园区的设计和建设应当避免过度奢华和资源浪费，要充分利用自然条件合理介入既有环境并做到最优，避免设计用力过猛产生破坏性的影响。

AT： 您谈到了有些高科技园区已经对城市的结构等方面产生了重大影响，城市结构出现"两极化"，请您详细解读一下。

王振军： 北京某网站发布的交通潮汐的动画显示出车流聚集的地方。可以看到，除了CBD（国贸）以外，其他密集的区域都是科技园区，这就是现在的北京城区所呈现出来的空间结构特点。

从这个交通流动图我们可以分析出：后工业时代或者说信息时代，城市的结构其实已经开始呈现两极化——一端为CBD，另一端为科技园区。所谓的科技园区其实定义很广，在二、三线城市可能叫工业园或高科技园区；但在北京、上海等一线城市，这种知识型的园区则被称为科技园区或智慧园区。以北京为例，二十年前，精英人士讨论工作的地点基本离不开西单、王府井等城市核心区域，那时候分布在西单地区的都是非常好的工作单位，比如中国移动、各大银行总部等。但现在情况变了，大家约见的地点已经变成CBD、中关村等经济开发区了，这是当下不可否认的城市现象。这个现象背后的本质就是基于互联网的信息社会。除了城市结构的两极化，目前高科技园区面临的另外两个问题是：（1）建设量大，建设速度惊人（表1）；（2）相关总结不够，研究不深，标准欠缺。

AT： 现在的高科技园区建设，最重要的是找准园区的产业定位、方向以及招商对象。一般业主或建设方在一开始不会有明确的想法，这就需要建筑师协助甲方招商、策划部门做园区产业结构策划等规划设计前期的工作。请您谈谈前期规划设计的具体内容和方法。

王振军： 你谈到的这个问题，可以归纳为是高科技园区设计中最关键的定性设计和定量设计问题。

定性就是找准园区定位和方向。比如任务书中，业主要求研发部分的建筑面积是10万 m^2，在这其中未来入驻的大型企业、中型企业、小型企业的数量是多少，比例怎么分配就是定量设计。很显然，不同城市、不同区位的高科技园区对研发模块规模的需求量是不一样的，建筑师要把它们进行拆分，比如把10万 m^2 拆成10个单位，分别确定2万 m^2 的企业数量，4000~5000m^2 的企业数量，2000~3000m^2 的企业数量。只有在确定这些之后，才能保证设计有弹性，这也是定量设计的难点和重点。除非极特殊的情况，否则不会出现那种10万 m^2 均分为10栋建筑的情况。当然，这也勉强算是一种定量方式，但其弹性就非常小了。

定量设计并没有一个完全严谨的方法和一个绝对完善

生长与蔓延　蔓·设计实践 I

国家级软件产业基地概况　　　　表1

院区名称	中关村	上海	大连	西安	广州	长沙	江苏	山东	珠海	杭州	西部	天津	深圳	总计
规划面积（ha）	139	135.6	748	67	2943	265	612	650	34	38	73.3	35.3	27	5767.2
建筑面积（万m^2）	62	116.6	237	11	863	150	528	518	33.7	12.6	68	42.4	16	2658.3

注：数据信息由国家发改委、信息产业部、商务部共同认定。

的标准。我们设计过的很多园区，业主方多数都没有什么经验，甚至有些评审团的专家在这方面经验也不多，这时建筑师的园区设计经验和对相关产业特点的把握就尤为重要。前期策划阶段的依据非常少，比如招商方面，谁也不能确定园区建设完成后招商入驻的企业一定是哪些家，市场行为就是导致设计不定性的原因所在。如果建筑师设计的园区多了，接触的业主、招商团队和入驻使用方多了，就会有一定的经验。通俗地讲，这就是所谓的设计弹性，只有有了弹性才能保证园区的灵活性和建设的成功。同时，政府的政策也要重点考虑。对于招商目标，前期也要有一定的针对性，这样才能给后续的设计提供建议和参考。

举个例子，中关村某园区的前期规划设计非常好，但最后园区的运营方、使用者和入驻企业却对园区评价不高。因为园区的设计太过于理想化，把硅谷高科技产业区的模式整套移植过来，但硅谷的实际情况是人少、面积大、地价便宜（当时如此）。而中关村软件园建筑只能放置在一个个"小气泡"中，园区的容积率大概是0.3，规划设计也缺乏弹性，基本无法扩建。园区运营了很多年，一直处于亏损状态。这正是策划方和设计方没有考虑到未来的发展需求和建设成本造成的。后来，经过政府特批，园区的设计又报批过一次，重新修改了一些规划，才勉强解决了一些问题。

AT：在做好园区的定量设计后，还需要解决哪些问题？对于园区运营，建筑师能做哪些工作？

王振军：高科技园区建设的关键因素离不开政策支持、创新技术资源（软环境）、规划设计（物理空间及配套功能支持）、运营管理（维护发展）等。对于高科技园区设计，要解决的问题就是建筑师在综合业主的意愿和自己的判断之后，用城市规划、建筑学、景观设计的方法，将园区各要素进行定性和定量的有机组合，并确定其相对比例和相对位置，使园区空间结构和环境品质达到最佳状态，以保证园区高效、健康、可持续地运行。

高科技园区的建设已经成为一个普遍的城市现象，很多设计院都在做高科技园区的规划设计，但是真正进行系统研究的并不多。我本人很早就关注了这方面，带领设计团队和研究生做了相关的研究工作，并以高科技园区中的典型软件园为例编写了《软件园区规划设计规范》，这也是我国第一部关于园区的规范，但即使是这样，仍然感觉远远不够，未来需要做的工作还有很多。

关于园区运营的问题，可能建筑师话语权不多，重要的是帮助业主做好前期的定性规划和定量设计，可以用三个词来概括：市场化管理、企业化运营、专业化服务，能做到这三点，基本上园区的运营应该没有问题，上海浦东张江的科技园区在这方面普遍做得不错。

关于高科技园区的规划和设计问题，能够展开探讨的话题太多，但理论终归要回归到实践中去，下面就以本人实践的项目为例，详细介绍高科技园区设计中的问题及解决方法，希望对同行有所裨益。

1　上海浦东软件园一、二、三期

上海浦东软件园位于上海浦东张江高科技园区的中心地带。软件园一、二期与三期之间隔祖冲之路遥相呼应，设计遵循"蔓·设计"理念演绎出各自独具特色的园区空间和形式，前者为井然有序的学院派风格，后者则更加现代简洁、自由开放，二者建成时间相隔近十年。

软件园一、二期的长条形基地尺度（240m×500m）和软件研发建筑尺度使"环形车道＋中心庭院"的规划结构显得顺理成章，而庭院中职工餐厅的绿化屋顶与来自主入口方向小于5%的微地形景观有机地衔接在一起，形成了一个放大、立体化并富有情趣的公共生态休闲空间，确保环绕周围的研发建筑能够拥有优质的景观。主楼在入口处的二层架空处理，使城市空间、园区空间以及园区的核心中庭空间有机地联系在一起，既开放又富有层次。环形道路的交通系统有效地保证了园区中庭的安静和舒适。上述设计在有着24m限高、容积率大于1.5的园区空间中非常难得。

软件园三期的规划在基地四边道路入口位置不变的前提下，打破了原有控规的中心十字路规划结构，采用环形主路将基地划分成不同规模的地块，就近将吕家浜的河水引入基地中心，形成"环形主路+中心湖"的规划结构，有效地降低了驶入园区的车速，同时这种向心式的富有层次的布局能够为满足不同规模的企业入驻提供更加多样化的选择。中心湖区沿岸的公共性建筑以合理的覆盖半径为研发企业提供服务，中心湖也成为休憩的中心、交流的中心、研发成功时庆典的中心，整个规划生动地诠释出建筑师主张的设计价值观——"科技只是一种媒介，人与自然通过它有机地联系在一起"。

2 中国信达（合肥）灾备及后援基地

形体设计增加与中庭环境的接触面积。面向城市的一面继续采用连续界面的手法将建筑组群有机地整合在一起，而将基地核心——数据中心处理成具有地域特点的新徽派建筑风格，以体现文化传承和人文关怀；基地的制高点——研发大楼在向主干道方向张开的同时与远处的巢湖风景区作出呼应。建筑外立面材料选用温润柔和的陶土板搭配简洁大方的玻璃幕墙，形成天然与人工、传统与现代、感性与理性的对比，二者刚柔并济，共同塑造出建筑的外部表情。

3 海盐杭州湾智能制造创新中心

海盐杭州湾智能制造创新中心项目可能是中国距大海最近的园区，基地位于国家中欧合作试点的浙江省嘉兴市海盐县，临近滨海公园，距离杭州湾不到800m。规划以"交融"为概念，希望城市、海洋、人、科技与自然在此交织，最终交融在自然的脉络中。建筑高度和体量由北向南朝大海方向逐渐降低和缩小；环形主干道+组团支路的交通系统为中心景观湖实现纯步行方式提供了保障；单体建筑将交通及辅助空间进行模块化处理，与主研发楼体量组合在一起，为园区建设阶段的产业化提供可能。作为园区中心和制高点的总部大楼，是整个园区的标志性建筑，以V字形平面面向大海，以争取最大化的景观和观海面，塑造出北侧欢迎之态的主入口广场空间；空间和形式的处理则通过左右两个体量的错动、穿插、咬合等手法追求建筑的动态平衡和协调两个体量的关系，同时寄予"合作、信念、信心"等美好的期许。

本文引改自《建筑技艺》2017年第10期

The Reflections on the Planning, Design, and Construction of Hi-Tech Parks

China's first software park, the Shanghai Pudong Software Park (Phases I and II), was completed under the direction of Chief Architect Wang Zhenjun from the China Electronic Engineering Design Institute. Subsequently, he led his team to complete dozens of various high-tech parks, including software parks, across the country. In 2013, he served as the chief editor for China's first set of design standards for science and technology parks, the "Software Park Planning and Design Standards", which was officially published. Additionally, he summarized his experience in park design and authored over a dozen papers published in leading academic journals such as "Architectural Journal". In September 2017, Liu Xiaonan and Zhu Xiaolin from the "Architectural Techniques" (AT) magazine conducted an exclusive interview with him.

AT: In China, there has been a proliferation of industrial parks in large and medium-sized cities. How do you define a hi-tech park, and what characteristics does it possess?

Wang Zhenjun: The planning and construction of a park, regardless of whether it occupies 100,000m^2 or 1,000,000m^2, will have certain impacts on the surrounding city, such as the flow of people, traffic, municipal construction, urban spatial structure, and industrial patterns. Most importantly, as a new spatial form within a city, hi-tech parks have greatly influenced the urban structure. Hi-tech parks and central business district (CBD) have formed two poles of the city. Therefore, it is necessary for architects to pay closer attention and think about how to integrate innovative elements into the planning and design of hi-tech parks and truly bring them into the city.

Furthermore, the construction of hi-tech parks should not only focus on innovation but also fully consider and understand the local urban and cultural context. For example, designing hi-tech parks in Suzhou, Qingdao, and Harbin may have similar initial planning formats and models, but the final built outcomes will undoubtedly differ. Another example is the construction of the Beijing Sub-Center in Tongzhou District, where the Beijing Municipal Party Committee and Government are planning to relocate. This will require a new strategy. In my opinion, it is important to explore and study the relationship between humans and nature in park construction. This can be summarized by effectively managing the relationship between the following three elements: people, physical environment, and technology and production methods. In 1998, I published an article titled "Integrating Humans and Nature through Technology", which encapsulates this concept. Therefore, the architect's task is to integrate these three elements through architectural and landscape language. This is the fundamental problem and starting point that needs to be addressed. The design and construction of hi-tech parks should avoid excessive extravagance and resource wastage. Instead, they should make full use of natural conditions to integrate harmoniously with the existing environment and achieve optimal results, while avoiding excessive disruptions caused by design interventions.

AT: You mentioned that some hi-tech parks have already had a significant impact on urban structure, leading to a "polarization" of urban structure. Could you please elaborate on this?

Wang Zhenjun: The traffic congestion animation published by a Beijing website shows the areas where the traffic flow is concentrated. As can be seen, apart from the central business district (CBD), the other densely populated areas are all hi-tech parks. This is the spatial structure characteristic currently observed in Beijing's urban area.

From this traffic flow map, we can analyze that in the post-industrial or information age, the urban structure has started to exhibit polarization – one end being the CBD, and the other end being the hi-tech parks. The term "hi-tech park" has a broad definition. In second and third-tier cities, it may be referred to as an industrial park or hi-tech

Table 1 Overview of national software industrial parks

Hi-tech park names	Beijing Zhongguancun	Shanghai	Dalian	Xi'an	Guangzhou	Changsha	Jiangsu	Shandong	Zhuhai	Hangzhou	Western region	Tianjin	Shenzhen	Total
Planned area (ha)	139	135.6	748	67	2943	265	612	650	34	38	73.3	35.3	27	5767.2
Built area (10,000m^2)	62	116.6	237	11	863	150	528	518	33.7	12.6	68	42.4	16	2658.3

Note: The data information is jointly recognized by the National Development and Reform Commission, Ministry of Industry and Information Technology, and Ministry of Commerce.

park; but in first-tier cities like Beijing and Shanghai, these knowledge-based parks are referred to as hi-tech parks or smart parks. Taking Beijing as an example, twenty years ago, the preferred locations for elite professionals to work were mainly in the core areas of the city such as Xidan and Wangfujing. The companies located in the Xidan area were considered to be excellent, such as China Mobile, and the headquarters of major banks. But now, the situation has changed. The meeting locations have shifted to the CBD, Zhongguancun, and other economic development zones. This is an undeniable urban phenomenon. The underlying essence of this phenomenon is the information society based on the internet. In addition to the polarization of urban structure, hi-tech parks currently face two other issues: 1) large-scale construction with astonishing speed (Table 1); 2) insufficient related research and lack of standards.

AT: The construction of hi-tech parks nowadays is primarily focused on determining the industrial positioning, direction, and target investors for the park. Generally, at the beginning, the owners or developers may not have a clear idea, which requires the assistance of architects and planning departments in the pre-planning and design phase, such as assisting in investment attraction and planning the industrial structure of the park. Please discuss the specific content and methods of the pre-planning and design.

Wang Zhenjun: The issue you mentioned can be summarized as the most crucial qualitative and quantitative design problems in hi-tech park design.

Qualitative design involves determining the park's positioning and direction. For example, in the project brief, the owner may require a built area of 100,000m^2 for R&D purposes. Within this, the number and proportion of large, medium, and small enterprises to be accommodated in the future are part of the quantitative design. Obviously, the demand for R&D module size varies in different cities and locations. Architects need to break it down, for example, dividing the 100,000m^2 into 10 units and determining the number of enterprises with areas of 20,000m^2, 4,000-5,000m^2, and 2,000-3,000m^2, respectively. Only after determining these can design flexibility be ensured, which is also the difficulty and focus of quantitative design. Unless under extremely special circumstances, it is unlikely to have a situation where the 100,000m^2 is evenly divided into 10 buildings. Of course, this can be considered a kind of quantitative approach, but it lacks flexibility.

Quantitative design does not have a completely rigorous method or an absolutely perfect standard. In many of the parks we have designed, most owners have little experience in this field, and even some experts on evaluation committees have limited experience. At this time, the architect's experience in park design and understanding of relevant industry characteristics become particularly important. There is very little basis in the early planning stage. For example, in terms of investment attraction, no one can determine which companies will be attracted to the park after its construction is completed. Market behavior is the reason for the design's uncertainty. If architects design more parks and have more interactions with owners, investment attraction teams, and occupiers, they will gain more experience. In simpler terms, this is what we call design flexibility, which ensures the flexibility of the park and the success of its construction. At the same time, government policies must also be given special consideration. For investment attraction targets, there must be some specific targeting in the early stages, in order to provide recommendations and references for subsequent design.

For example, the initial planning and design of a certain park in Beijing's Zhongguancun area was very good, but the park's operators, users, and the companies that moved in later did not rate the park highly. This was because the design of the park was too idealized, trying to transplant the model of Silicon Valley's high-tech industrial area as a whole. However, the actual situation in Silicon Valley was characterized by a small population, a large area, and low land prices (at that time). In contrast, the buildings in Zhongguancun Software Park were only placed in individual "bubbles", with a plot ratio of about 0.3, and the planning design lacked flexibility, making it difficult for the park to expand. The park operated at a loss for many years. This was precisely caused by the failure of the planning and design teams to consider future development needs and construction costs. Later, with special approval from the government, the park's design was re-approved and some planning was modified, which barely solved some of the problems.

AT: After completing the quantitative design of the park, what other issues need to be addressed? What can architects do in terms of park operation?

Wang Zhenjun: The key factors in the construction of hi-tech parks include policy support, innovative technological resources (soft environment), planning and design (physical space and supporting functions), and operation management (maintenance and development). In terms of hi-tech park design, the problem to be solved is how architects, after considering the comprehensive intentions of the owners and their own judgments, can use methods from urban planning, architecture, and landscape design to qualitatively and quantitatively combine various elements of the park, determining their relative proportions and positions. This aims to achieve the optimal spatial structure and environmental quality of the park, ensuring its efficient, healthy, and sustainable operation.

The construction of hi-tech parks has become a common urban phenomenon, and many design institutes are involved in planning and designing hi-tech parks. However, there are not many who actually conduct systematic research in this area. I have personally been paying attention to this for a long time and have led design teams and graduate students to conduct related research. I have also written the "Planning and Design Standards for Software Parks" based on typical software parks in hi-tech parks. This is the first standard on parks in China.However, even with that, I still feel that it is far from enough, and there is still much work to be done in the future.

Regarding park operation, architects may have limited influence. The important thing is to assist the owners in conducting qualitative planning and quantitative design in the early stages. This can be summarized in three aspects: market-oriented management, enterprise-oriented operation, and professional services. By achieving these three aspects, the park's operation should be generally problem-free. The Zhangjiang Hi-Tech Park in Shanghai's Pudong district has done well in this regard.

There are many topics that can be discussed regarding the planning and design of hi-tech parks. However, theory ultimately needs to return to practice. Therefore, I will use my own practical projects as examples to provide a detailed introduction to the issues and solutions in hi-tech park design. I hope it will be beneficial to colleagues in the field.

1 Shanghai Pudong Software Park Phase I ,II ,and III

Shanghai Pudong Software Park is located in the central area of Zhangjiang High-Tech Park in Pudong,Shanghai. Phases I and II of the software park are separated by Zuchongzhi Road from Phase III,complementing each other. The design follows the "*Organically-Permeated Design*" concept,interpreting unique spatial and formal characteristics for each phase. The former (Phases I and II) features an orderly, academic style, while the latter (Phase III) is more modern, concise, free, and open. The construction of the three phases was nearly a decade apart.

The long and narrow site scale (240m×500m) of Phase I and II of the software park, along with the scale of the software research and development buildings, makes the planning structure of "ring road + central courtyard" appear logical. The green roof of the staff canteen in the courtyard is seamlessly connected to the micro-terrain landscape from the main entrance, forming an enlarged, three-dimensional, and intriguing public ecological leisure space, ensuring that the surrounding research and development buildings have high-quality landscapes. The main building is elevated on the second floor at the entrance, creating an organic connection between urban space, park space, and the core atrium space of the park, which is both open and hierarchical. The traffic system of the ring road effectively ensures the tranquility and comfort of the park's atrium. The aforementioned design is rare in a park space with a height limit of 24m and a plot ratio greater than 1.5.

In the planning of Phase III of the software park, while keeping the entrance positions of the roads around the site unchanged, the original central crossroads planning structure of the control plan is broken. An annular main road is used to divide the site into different-sized plots, and the river water from Lujia Bang is introduced into the center of the site, forming a planning structure of "annular main road + central lake". This effectively reduces the vehicle speed when entering the park. At the same time, this concentric and hierarchical layout provides more diversified choices for enterprises of different scales. Public buildings along the shores of the central lake provide services to research and development enterprises with a reasonable coverage radius. The central lake has also become the center of relaxation, communication, and celebration of research and development achievements. The entire planning vividly demonstrates the architect's design values of "technology is just a medium, through which people and nature are organically connected".

2 China Xinda (Hefei) Disaster Recovery and Back-up Base

The design of the architectural form elements increases the contact area with the courtyard environment. The side facing the city continues to integrate the architectural clusters organically through continuous interfaces, while the core of the base, the data center, is designed in a new Huizhou architectural style with regional characteristics to reflect cultural heritage and humanistic care. The research and development building, the highest point of the base, opens towards the main road while responding to the distant Chaohu Scenic Area. The exterior facade of the buildings combines warm and gentle terracotta panels with sleek and elegant glass curtain walls, creating a contrast between nature and artificiality, tradition and modernity, sensibility and rationality. The combination of these elements shapes the external expression of the architecture, harmonizing both strength and flexibility.

3 Haiyan Hangzhou Bay Intelligent Manufacturing Innovation Center

The Haiyan Hangzhou Bay Intelligent Manufacturing Innovation Center project is possibly the park closest to the sea in China. The site is located in Haiyan County, Jiaxing City, Zhejiang Province, which is part of the national China-Europe cooperation pilot zone. It is adjacent to the coastal park and less than 800m away from Hangzhou Bay. The concept of "integration" is the guiding principle of the planning, aiming to intertwine the city, the ocean, people, technology, and nature, ultimately merging them within the context of nature. The height and volume of the buildings gradually decrease and shrink from north to south, facing the sea. The transportation system, consisting of a circular main road and cluster roads, guarantees a pedestrian-oriented environment around the central landscape lake. The individual buildings are modularized in terms of transportation and auxiliary spaces and combined with the main research and development building, providing possibilities for the industrialization stage of park construction. As the headquarters building, which serves as the center and the highest point of the park, it is the iconic structure of the entire park. Its V-shaped plan faces the sea to maximize the scenic and sea views, creating a welcoming gesture with the main entrance plaza on the north side. The spatial and formal treatment pursues dynamic balance and coordination between the two volumes through techniques such as shifting, interlocking, and biting, while embodying the beautiful expectations of "cooperation, belief, and confidence".

This article is adapted from *Architectural Technique*, Issue No.10, 2017.

设计实践

已建项目：72页–251页
在建项目：252页–275页
未建项目：276页–287页

Design Practices

Completed Projects: p.72-251
Projects Under Construction: p.252-275
Unbuilt Projects: p.276-287

上海浦东软件园一期、二期工程 · Pudong Software Park, Phase I, II In Shanghai

设计 Design 1995 / 1997 · 竣工 Completion 2000 / 2003

地点：上海 · 用地面积：3.00 公顷 / 7.71 公顷 · 建筑面积：3.11 万平方米 /9.24 万平方米
Location：Shanghai · Site Area：3.00ha/7.71ha · Floor Area：31,100m² /92,400m²

合作设计师：郭应平、李华、邓涛、李娜、韩合军、郭珍珍、李小华、姜志勇、王峰
Co-designer：Guo Yingping, Li Hua, Deng Tao, Li Na, Han Hejun, Guo Zhenzhen, Li Xiaohua, Jiang Zhiyong, Wang Feng

　　上海浦东软件园位于上海浦东张江高科技园区的中心地带。软件园一、二期与三期之间由祖冲之路贯穿，二者遥相呼应而又各自成一体，将"蔓·设计"理念演绎出各自独具特色的空间形态，前者为井然有序的学院派风格，后者则更现代简洁、自由开放。两处园区前后相距十年建成。

　　软件园一、二期的长条形基地尺度（240m×500m）和软件研发建筑尺度使环形车道＋中心庭院的规划结构显得顺理成章，而庭院中职工餐厅的绿化屋顶与来自主入口方向小于5%的微地形景观有机地衔接在一起，形成了一个被放大、被立体化的并富有情趣的公共生态休闲空间，优质的景观可被环绕周围的研发建筑中的人们一览无余。主楼入口处的二层架空处理，使城市空间、园区空间以及园区的核心中庭空间有机地联系在一起，既开放又富有层次。环形道路的交通系统有效地保证了园区中庭的安静和舒适。以上这些在有着24m限高、容积率大于1.5的园区空间中显得非常难得和珍贵。陶土面砖与乳白色涂料的搭配更彰显了园区返朴归真的气质。

Shanghai Pudong Software Park is located in the central area of Zhangjiang Hi-Tech Park in Pudong, Shanghai. Separated by Zhuchongzhi Road, Phase I and II of the Software Park echo with Phase III at a distance, yet each having its own features; the two parts interprets the "organically-permeated design" concept in distinct spatial forms; the former presents itself in an orderly academic style, while the latter is more modern, concise, free and open. The completion of the two parks are almost 10 years apart.

Based on the rectangular site (240m×500m) of Phase I and II of the Software Park and the dimensions of the software R&D buildings, the planning structure of a circular driveway plus a central yard is logical. The green roof of the staff cafeteria in the yard is organically linked with the micro-relief landscape with a slope of less than 5% from the main entrance, thus forming a public ecological leisure space that is magnified, three-dimensional and enjoyable. The superior landscape is within the field of view of the people who work in the ambient R&D buildings. The open space in the second floor over the entrance of the main building connects the urban space, park space with the core atrium space in the park, creating open and tiered space for the park. The road system of a traffic circle effectively ensures the quiet and comfortable atmosphere for the park atrium, which seems very difficult and valuable for the park space that has a 24m height limit and floor area ratio of greater than 1.5. The combination of terracotta tiles and oyster white paint further emphasizes the park's return-to-nature aesthetic.

一、二期模型 / Model of Phase I and II
设计草图 / Sketch

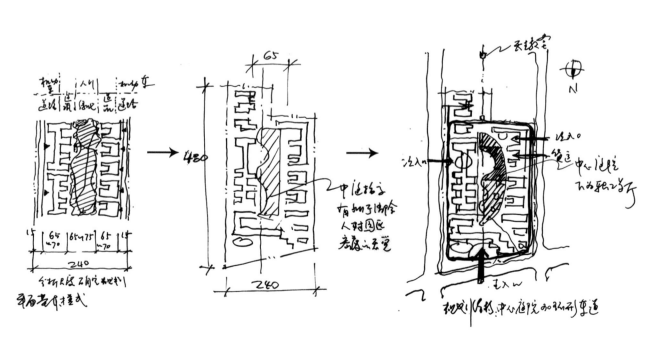

总平面 / General Layout

上海浦东软件园一期、二期工程

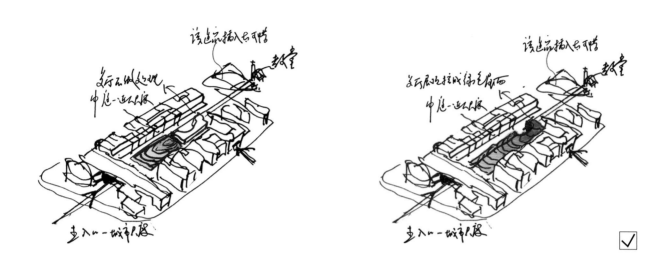

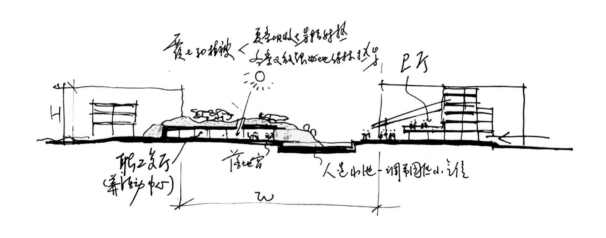

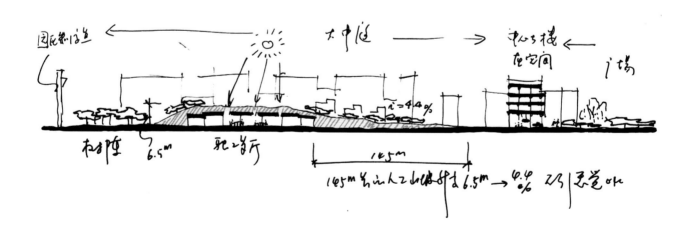

设计草图 / Sketch

首都国际机场东区塔台 · Control Tower of Beijing Capital International Airport East District

设计 Design 2004 · 竣工 Completion 2007

地点：北京 · 用地面积：1.33 公顷 · 建筑面积：2647 平方米
Location：Beijing · Site Area：1.33ha · Floor Area：2,647m²

合作设计师：刘嘉嘉、张会明、李娜、董昭英、邓涛、郭珍珍、王凯、姜志勇、王峰
Co-designer：Liu Jiajia, Zhang Huiming, Li Na, Dong Zhaoying, Deng Tao, Guo Zhenzhen, Wang Kai, Jiang Zhiyong, Wang Feng

城市的机场已是现代城市生活越来越重要的节点之一，导航塔台作为机场的指挥枢纽和视觉核心，自然成为一个绕不开的建筑类型，会因成为机场及其所在城市的标志物而备受关注。在充分满足塔台主要功能需求前提下，协调其与航站楼的关系，以及对于地域、文化、情感所形成的精神和心理意义的诠释，使塔台"介入"城市后能够对当下和未来产生可持续的积极影响，就成为另一重要任务。东区塔台位于T3航管楼北端。塔台的构思源于对先期确定的、由两个Y字对接在一起的T3航站楼方案的研究。将航站楼平面立体化形成了塔台基本形态，自然地使两个尺度悬殊的建筑取得了呼应，而这一形态又真实地反映了现代导航塔台的功能本质，外装材料除塔身拟采用环氧型外墙涂料外，其余部位为复合铝板幕墙。材料的搭配充分考虑了整个建筑的虚实对比、质感对比以及维护、清洗等方面因素。竖向的金属立挺使塔台向上的张力得到加强，弥补了82m限高带来的遗憾。东区塔台在满足视线无障碍的前提下，采用国际先进的集约型导航明室型制，降低了塔台的造价。

Urban airports rely heavily on their control towers, which serve as both command centers and visual landmarks.These towers are crucial not only for their primary functions but also for harmonizing with terminal buildings and reflecting the region's cultural and emotional significance.The East Control Tower, located at the northern end of the T3 Air Traffic Control Building, is designed to complement the T3 terminal layout.Its unique Y-shaped structure establishes a dialogue with the terminal, showcasing modern navigation control tower functionality.The exterior materials, except for the tower body which uses epoxy-type exterior wall paint, consist of composite aluminum panel curtain walls for the other parts. The material selection fully considers the contrasts in the building's solidity and transparency, texture differences, as well as factors related to maintenance and cleaning. The vertical metal fins enhance the upward tension of the tower, compensating for the limitations imposed by the 82-meter height restriction. The T3 control tower employs an internationally advanced compact navigation light system, ensuring unobstructed sightlines while also reducing the overall construction cost of the tower.

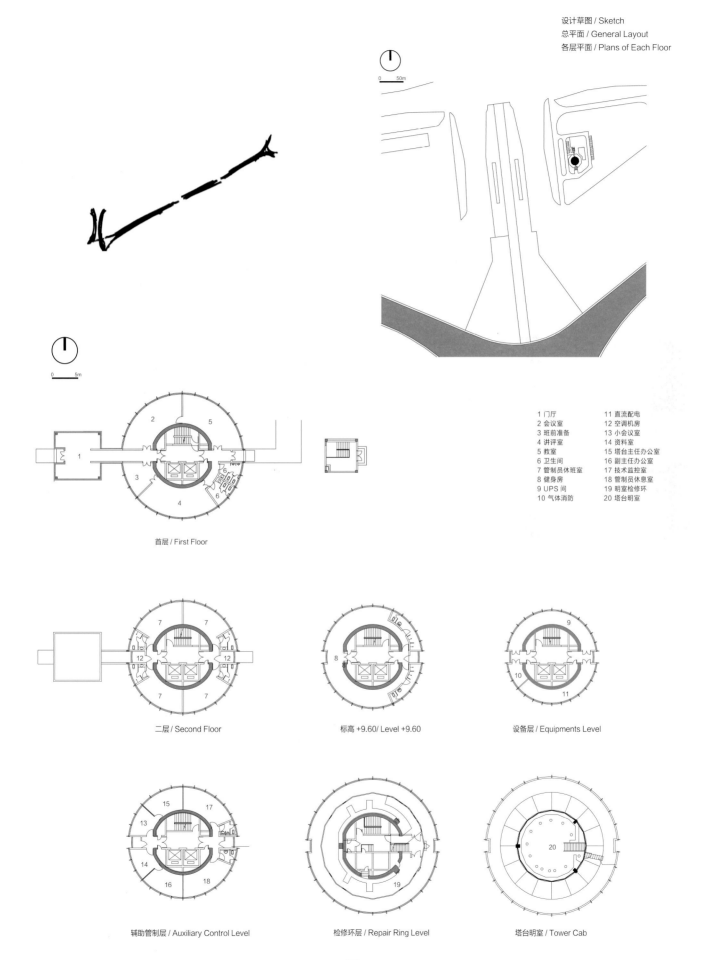

生长与蔓延　蔓·设计实践 I

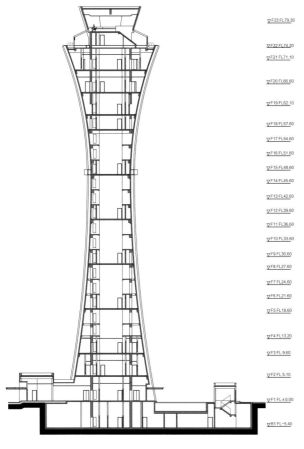

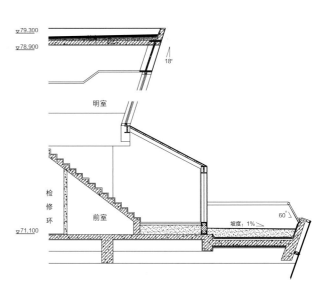

明室详图 / Details
剖面 / Section

北京国际财源中心（IFC）· Beijing International Finance Center

设计 Design 2004 · 竣工 Completion 2010

地点：北京 · 用地面积：2.03 公顷 · 建筑面积：26.67 万平方米
Location: Beijing · Site Area: 2.03ha · Floor Area: 266,700m²

合作设计师：高志、刘震宇、娄宇、闫谧、姜志勇、李娜、李华、矫金广、郭珍珍、王凯
Co-designer: Gao Zhi, Liu Zhenyu, Lou Yu, Yan Mi, Jiang Zhiyong, Li Na, Li Hua, Jiao Jinguang, Guo Zhenzhen, Wang Kai

位于北京国贸 CBD 商圈的基地沿长安街呈长条形（280m×65m），并跨街区展开。方案采用模块化（54m×28m）手法，在横竖两个维度进行模块组合后沿长安街布置，建筑顶部采用"世界之窗"的设计使两个模块更具统一性；建筑用极简手法使之在 CBD 林立的建筑群体中脱颖而出。两高两低的塔群造型完美地融入了长安街起伏优美的天际轮廓中，同时高低不同的塔楼为不同的使用者提供了规模灵活可选的 5A 级办公单元。单体标准层运用大跨度、窄核心筒和竖向穿插空中庭院等手法，意在为金融从业者带来更加舒适的工作体验。国际财源中心在设计中有效地减少对环境和使用者的负面影响，表现出了对使用者、投资者的关怀和社会责任。而且，绿色建筑也将为使用者带来营运成本的节约、员工工作效率的提高，因为人们总是倾向于在绿色建筑中工作。由于在总图、建筑、给水排水、暖通空调等专业中恰如其分的选用了节能技术，本项目荣获了美国 LEED 体系认证。

Located in the Beijing Central Business District (CBD), the site stretches along Chang'an Avenue in a rectangular shape (280m×65m) and spans across multiple blocks.The design employs a modular approach (54m×28m) for both horizontal and vertical dimensions, arranging along Chang'an Avenue, with the architectural roof adopting the design motif of the "Window of the World" to enhance unity between the two modules.Employing a minimalist approach, the structure stands out amidst the densely populated skyline of the CBD. The design of the two high and two low tower cluster seamlessly integrates into the picturesque skyline of Chang'an Avenue, while offering scale-flexible 5A-grade office units for different users.Standard floors feature large spans, narrow core shafts, and vertically interspersed atriums, aimed at providing a more comfortable working experience for financial professionals.The International Financial Center effectively minimizes negative impacts on the environment and users, demonstrating care for users, investors, and social responsibility.Moreover, green building practices are expected to bring cost savings in operations and enhance employee efficiency, as individuals tend to prefer working in green buildings.By judiciously incorporating energy-saving technologies in various aspects such as overall planning, architecture, water supply and drainage, and HVAC systems,the project has been awarded certification under the U.S. LEED system.

生长与蔓延　蔓·设计实践 I

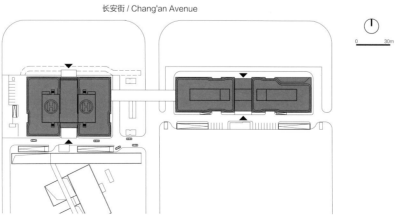

长安街 / Chang'an Avenue

总平面 / General Layout

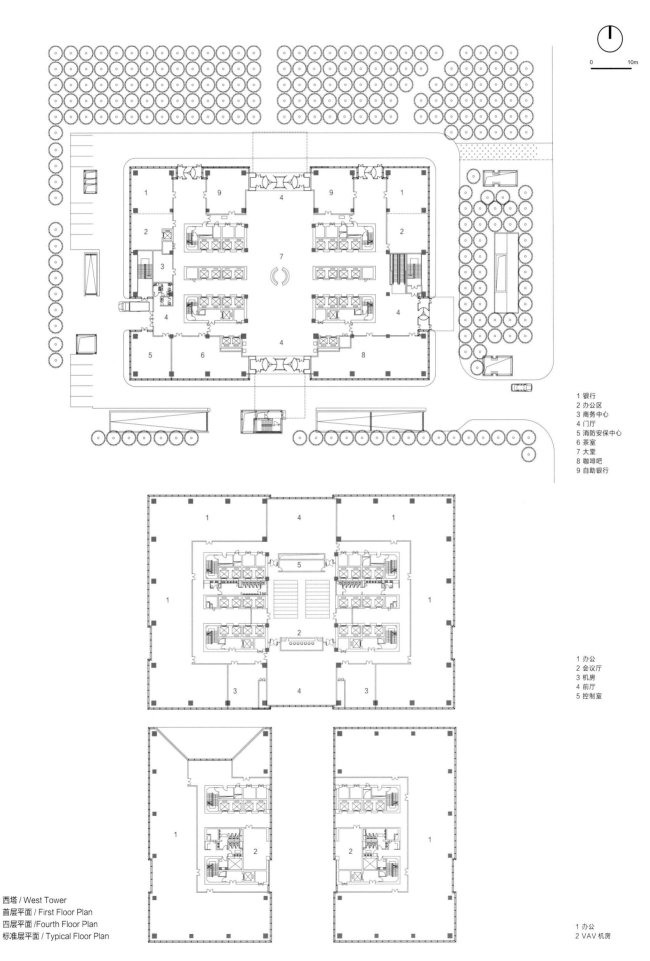

北京国际财源中心（IFC）

1 银行
2 办公区
3 商务中心
4 门厅
5 消防安保中心
6 茶室
7 大堂
8 咖啡吧
9 自助银行

1 办公
2 会议厅
3 机房
4 前厅
5 控制室

西塔 / West Tower
首层平面 / First Floor Plan
四层平面 / Fourth Floor Plan
标准层平面 / Typical Floor Plan

1 办公
2 VAV 机房

北京国际财源中心（IFC）

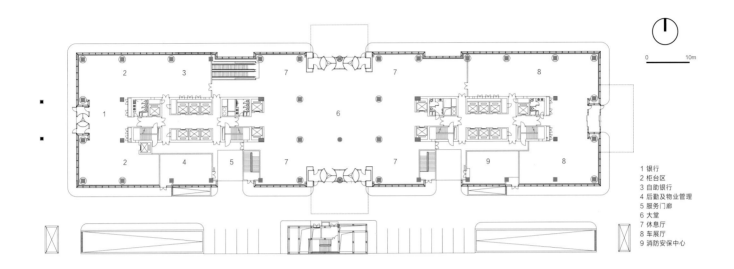

1 银行
2 柜台区
3 自助银行
4 后勤及物业管理
5 服务门廊
6 大堂
7 休息厅
8 车展厅
9 消防安保中心

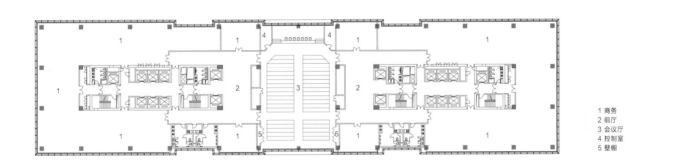

1 商务
2 前厅
3 会议厅
4 控制室
5 壁橱

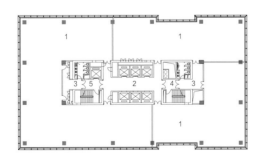 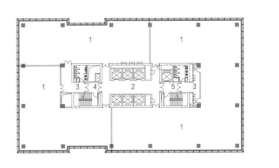

1 办公
2 电梯厅
3 走廊
4 前室
5 合用前室

东塔 / East Tower
首层平面 / First Floor Plan
三层平面 / Third Floor Plan
标准层平面 / Typical Floor Plan

上海国家软件出口基地（浦东软件园三期）· Shanghai National Software Export Base (Pudong Software Park Phase III)

设计 Design 2005 · 竣工 Completion 2012

地点：上海 · 用地面积：44.70 公顷 · 建筑面积：50.47 万平方米
Location：Shanghai · Site Area：44.70ha · Floor Area：504,700m²

合作设计师：张会明、权薇、朱谞、孙成伟、邓涛、李华、郭珍珍、王凯、姜志勇、王峰
Co-designer：Zhang Huiming, Quan Wei, Zhu Xu, Sun Chengwei, Deng Tao, Li Hua, Guo Zhenzhen, Wang Kai, Jiang Zhiyong, Wang Feng

　　软件园 III 期的规划在延续基地四边道路入口位置不变的前提下，打破了原有控规的中心十字路网规划结构，采用环形主路将基地划分成不同层级和不同规模的地块，就近将吕家浜的河水引入基地中心，形成环形主路加中心湖的规划结构。环形路一方面有效地减少了过境交通，降低了驶入园区的车速，同时这种向心式的富有层次的布局能够为满足不同规模的企业入驻提供更加多样化的选择。中心湖区沿岸的公共性建筑以极为合理的半径为研发企业提供着服务，中心湖成为了休憩的中心、交流的中心、研发成功时庆典的中心，整个规划生动地诠释了"科技只是一种媒介，人与自然通过它而有机地联系在一起"的设计价值观。

While keeping the position of the access to the ambient roads intact, the planning for Phase III of the Software Park divides the site into blocks of different tiers and various dimensions with the circular main road. Water from the Liujiabang Creek nearby is introduced into the center of the site so that the planning structure is characterized by the circular main road and the central pond. On the one hand, this structure effectively reduces the speed of vehicles driving into the park; on the other hand, the centripetal and hierarchical layout provides diversified options for resident companies of various scales. The public buildings along the bank of the central pond provide services to the R&D businesses with a very reasonable radius. The central pond becomes a center of repose, a center of communication, and a center of celebration for the success of R&D projects. The planning gives a vivid interpretation of the design values that "science and technology is a mere medium, through which people and nature are organically connected."

生长与蔓延　蔓·设计实践 I

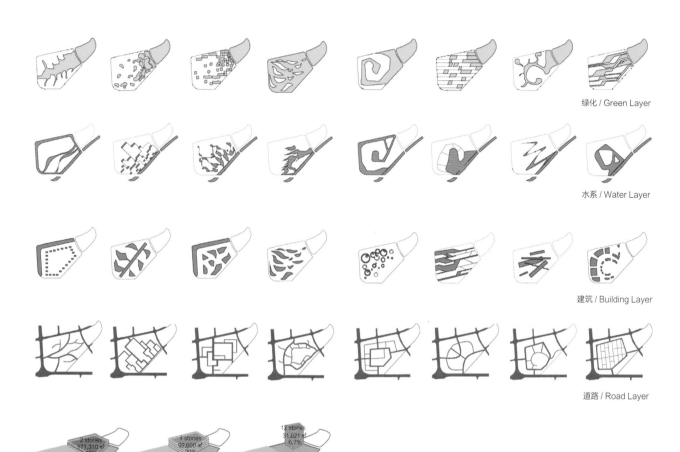

绿化 / Green Layer

水系 / Water Layer

建筑 / Building Layer

道路 / Road Layer

园区规划要素提取及研究 / Extraction and Study of Park Planning Elements
土地利用分析 / Land Ultilization Analysis

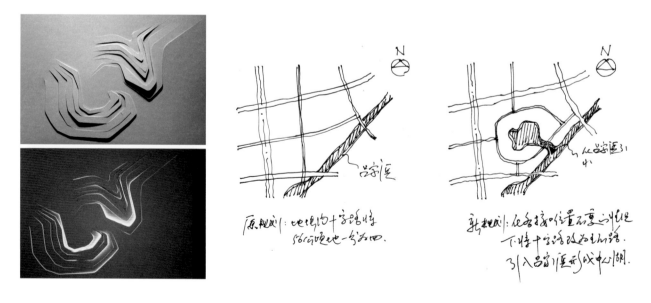

自然场力概念 / Concept of Natural Field Force
设计草图 / Sketches

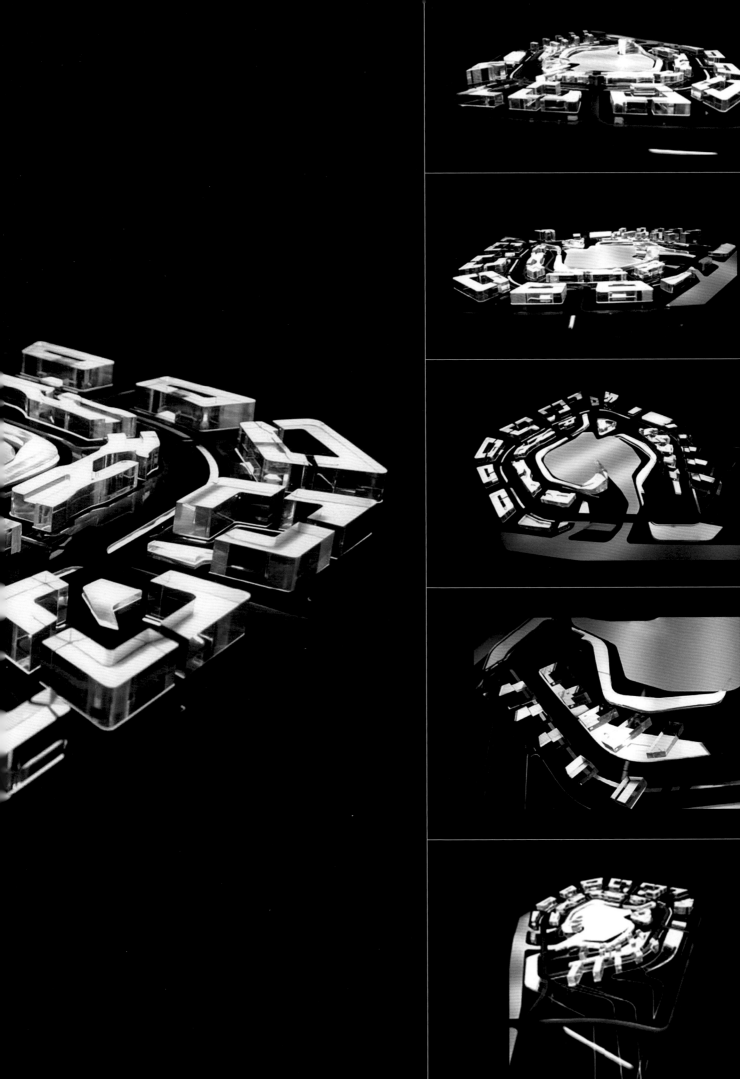

生长与蔓延　蔓·设计实践 I

研发

上海国家软件出口基地（浦东软件园三期）

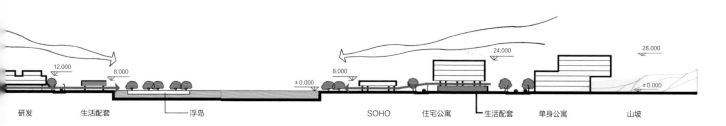

园区断面分析 / Section Analysis

2010年上海世博会沙特国家馆·The Saudi Arabia Pavilion at Shanghai Expo 2010

设计 Design 2007·竣工 Completion 2010

地点：上海·用地面积：6000 平方米·建筑面积：6126 平方米
Location: Shanghai · Site Area: 6,000m² · Floor Area: 6,126m²

合作设计师：张会明、孙成伟、邓涛、董召英、李华、朱谞、权薇、李娜、郭珍珍、王凯、李小华、姜志勇、王峰
Co-designer: Zhang Huiming, Sun Chengwei, Deng Tao, Dong Zhaoying, Li Hua, Zhu Xu, Quan Wei, Li Na, Guo Zhenzhen, Wang Kai, Li Xiaohua, Jiang Zhiyong, Wang Feng

沙特馆是2010年上海世博会上唯一由中国人独立设计完成的外国馆。该设计用一种轻松的建筑语言描绘了阿拉伯神话中的月亮船沿着海上丝绸之路从阿拉伯半岛漂浮到东方时尚港口——上海的场景。极具动感和未来感的体量漂浮在地表之上，上扩下收的造型营造出大量凉爽舒适的室外等候空间和表演空间；室内外参观流线围绕中庭环形布置，外环上行、内环下行的安排使进出人流舒缓有序，并且核心中庭充分利用自然采光和通风使建筑自成体系，达到节能环保的效果；空间展示上，建筑创新性地采用全球独创的"全景融入式立体参观方式"，即以船体内壳作为展示投影屏幕，展廊架设其上，用极具体验性的融入式动线布局，使其展示效果和文化信息传递量最大化，似驾乘在阿拉伯魔毯上的参观体验，并与等候区阿拉伯风情的景观、开敞的歌舞表演平台以及视野极佳的屋顶花园，一起为游客奉献了一场汇集阿拉伯人文风情与地域文化的饕餮盛宴。沙特馆以清晰的创作概念、建筑形象以及与形象高度融合的参观方式获得了超高人气，很好地诠释了有机建构理念的真正含义和价值所在。

The Saudi Arabia Pavilion at Shanghai EXPO 2010 is the sole foreign pavilion whose design is independently performed by a Chinese architect.The design depicts a scene of the moon boat in the Arabian mythology floating to the fashionable oriental port—Shanghai along the maritime Silk Road from the Arabia Peninsula by using a light-hearted architectural language.The extremely dynamic and futuristic mass stands adrift on the surface, and the V-shaped form creates a large area of cool and comfortable outdoor waiting space and performing space.The indoor and outdoor visitors' circulations are arranged around the atrium, giving ease and order to both the ascending visitor flow on the outer ring and the descending one on the inner ring. Besides, the natural lighting and ventilation introduced in the atrium enables the pavilion to establish a system of its own, achieving the effect of energy saving and environmental protection.In terms of spatial presentation, the project innovatively adopts a globally unique "panoramic immersive three-dimensional viewing approach." By utilizing the inner shell of the ship as a projection screen and suspending the exhibition gallery above it, the layout creates an immersive and experiential path. This maximizes both the exhibition's visual impact and the transmission of cultural information, offering visitors a sensation akin to riding an Arabian magic carpet. This experience, combined with the Arabian-themed landscape of the waiting area, an open dance performance platform, and a rooftop garden with exceptional views, delivers a rich feast of Arabian cultural and regional charm to visitors. The Saudi Pavilion has garnered immense popularity due to its clear creative concept, architectural imagery, and a visiting experience that is highly integrated with its overall design. It effectively embodies the true meaning and value of the organic construction philosophy.

生长与蔓延　蔓·设计实践 I

首层平面 / First Floor Plan
三层平面 / Third Floor Plan

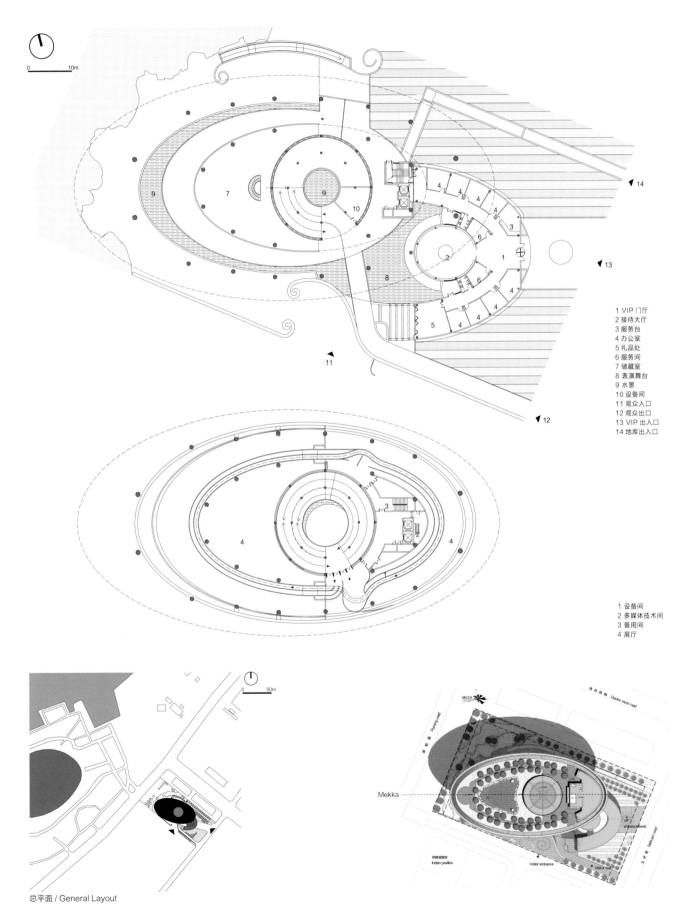

1 VIP 门厅
2 接待大厅
3 服务台
4 办公室
5 礼品处
6 服务间
7 储藏室
8 表演舞台
9 水景
10 设备间
11 观众入口
12 观众出口
13 VIP 出入口
14 地库出入口

1 设备间
2 多媒体技术间
3 备用间
4 展厅

总平面 / General Layout

2010年上海世博会沙特国家馆

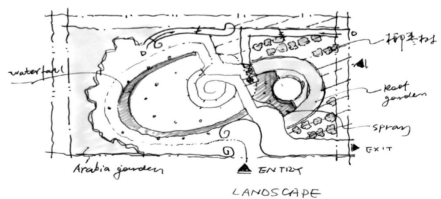

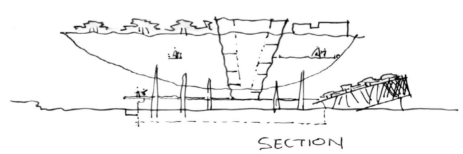

设计草图 / Sketch

传统：看与被看二元并置

创新：全景融入式观影模式

设计草图 / Sketch

2010年上海世博会沙特国家馆

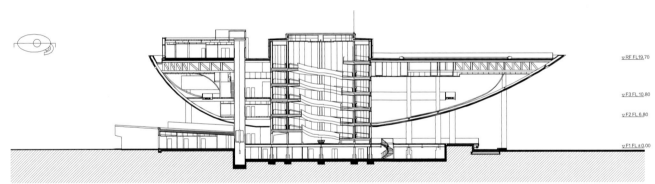

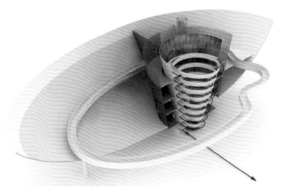

剖面 / Section
全景融入式参观流线 / Panoramic Integrated Visiting Route

105

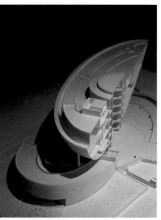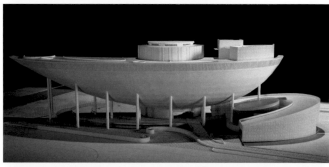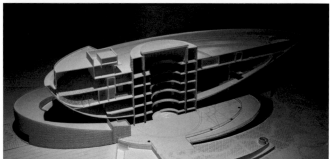

模型 / Mode
建筑表皮与灯光效果推敲 / Deliberation on Building Skins and Lighting Effect

2010年上海世博会沙特国家馆

生长与蔓延　蔓·设计实践 I

108

2010年上海世博会沙特国家馆

来自：穆罕默德·阿恩海姆迪教授
职衔：沙特馆执行官；沙特阿拉伯国王大学建筑系教授

From: Dr. Mohammad Alissan Alghamdi
Title: Executive directior of Saudi Pavilion;
Professor of Architecture and Building Science at King Saud University

来自：托马斯·赫斯维克
职衔：托马斯·赫斯维克工作室创始人、主持人；上海世博会英国馆建筑师

From: Thomas Heatherwick
Title: Founder and Director of Heatherwick Studio;
Architect of UK Pavilion at Shanghai EXPO 2010

来自：克利福德·皮尔逊
职衔：《建筑记录》杂志副主编

From: Clifford Pearson
Title: Deputy Editor of Architectural Record

- 我们最终选择了王先生的方案，因为该设计运用现代建筑语言、技术和材料出色地诠释了阿拉伯文化和传统，因而在本次设计竞赛中脱颖而出，斩获竞赛头奖。
  ——穆罕默德·阿恩海姆迪教授

We ultimately accepted Mr.Wang's proposal because he was the first place winner of this competition as he excelled in his proposal in reflecting the Arab culture and traditions with modern archi–language, technology and materials.
—— Dr. Mohammad Alissan Alghamdi

- 从世博会访客和中外建筑界的评价来看，沙特馆无疑是成功的，这要归功于设计师的工作和付出。这件作品展示出的建筑造诣使其成为当今中国建筑师中的佼佼者。
  ——托马斯·赫斯维克

The success of the Saudi Arabia Pavilion both with visitors to the Expo and the architectural community in China and overseas is a credit to his work and commitment. In his work he has risen to the level of accomplishment that is shared only by a few Chinese architect working today.
—— Thomas Heatherwick

- 我对于王振军先生将大胆的标志性结构与全方位多媒体室内展示高度融合的体验印象深刻。这个项目的成功使他跻身于世界级资质的精英建筑师队伍。沙特馆不仅成为了世博会最受游客欢迎的展馆之一，也在建筑师界及中外媒体的关注报道中广受好评。
  ——克利福德·皮尔逊

I was impressed by Mr.Wang Zhenjun's ability to integrate a bold iconic structure with a sophisticated multi–media interior experience. His success with this project places him in an elite group of architects with world–class credentials. Not only has the Saudi Arabia Pavilion became one of the most popular pavilions with visitors to the Expo, but it has garnered acclaim from architects and drawn the attention of publication in China and abroad.
——Clifford Pearson

国际同行评价 / Reviews from international peers

建筑秀场上的文化容器——沙特馆

王振军　张会明　董召英　孙成伟

沙特阿拉伯国家馆（简称沙特馆）是上海世博会外国自建馆中唯一由中国设计团队原创设计的场馆。建筑师认为世博会国家馆的功能首当其冲就应是浓缩各个国家文化和传统的载体，概念的产生应是在建筑师分析加感知这些文化背景资料并结合建筑学的知识整合的结果。

1 寻找与文化的对应关系

展馆设计理念取材于阿拉伯神话故事中的"月亮船"——它代表着幸福、美好。同时也是中沙海上丝绸之路物质与文化的载体。"月亮船"朝向伊斯兰圣城麦加方向（西偏北31.748°），给人极强的精神寄托感，这一具有精神含义的轴线转动为长方形用地外部空间营造和布局带来了契机。建筑周围用"沙漠""绿洲""海洋"组成沙特地域风貌，反映出沙特特有的人文风情和地域特征。

2 空间营造与场所精神

（1）等候空间——灰空间

从西南侧公共广场进入沙特馆入口广场，通过环形坡道按照穆斯林朝圣的顺时针方向依次排队等候，这时在公共平台上展演的民族歌舞可缓解等待带来的枯燥，从而确保等待过程也极富观赏性。结合景观在表演平台前方设置台阶看台，充分地考虑了各方位角度观赏效果，确保各区域游客的欣赏视线达到最佳。

（2）室内展示空间——多媒体技术和弧形自动步道技术支持下的全景融入式展示空间

参观过灰空间的歌舞表演，沿上行坡道依次进入室内展示空间。根据月亮船的总体概念，空间展示上我们创新性地利用船体内壳作为展示投影屏幕，展廊架空其上，环绕而行，用创新性的全景立体融入式动线布局，观众好似乘坐"阿拉伯魔毯"穿行其间，创造了独特的室内参观体验。多媒体技术（IMAX 3D）的发展又为这种畅想提供了强力的保证，使这一文化容器的展示效果和信息量达到最大化。

（3）屋顶阿拉伯花园——第二标高城市展示空间

一个极富异域特色的观光休息平台位于展馆顶层，不但增加了展馆空间情趣，又恰到好处地补充了这个区域登高望远的功能。整个展馆空间设计从城市空间—场地空间—中介空间—室内空间—屋顶开敞空间都充分强调层次感和场所感的塑造，畅游之中的人们在城市与建筑的内外空间转换中享受着沙特文化带来的浪漫体验。沿内环下行坡道进入1.0m平台出口处是纪念品店和咖啡厅，游客在此购买特色纪念品的同时可以品尝沙特的美味，由此结束本次参观。

场地及建筑在东侧设有VIP出入口，VIP客人和残障人士可在接待厅内乘观光（无障碍）电梯直接进入场馆进行参观，从而避免了与普通游客流线交叉，普通进出参观流线围绕中庭环形布置，人流有序分离。

3 节能环保设计

在建筑物本身设计中通过如下方法提高建筑对自然资源的充分利用：

（1）展馆内作为交通核心的中庭充分利用自然光，减少了常见展馆对人工照明过度依赖而导致的能耗。展馆外墙根据功能需要不设外窗，大大减少了室内外热交换的可能。

（2）用架空建筑体量来营造凉爽舒适的室外等候空间和表演空间。上海属亚热带季风气候，5~10月的世博会期间正值日照峰值段。场馆的架空结构在西北侧形成巨大的阴影区，为等候参观和观赏演出的人们提供了舒适的室外小环境。展示平台下侧周边的喷水设施又对空间起到了降温作用。

（3）船形展馆的上扩下收的造型，使展馆外墙几乎回避了日晒的影响，外围护隔热问题通过造型已初步解决。

（4）屋顶绿化在展示沙特风貌的同时又有效地改善了展厅屋面的隔热性能。

（5）充分利用自然通风及其压力、温差的作用，中庭空间促进了空气的流通，也有效地改善了展馆内的空气质量。

4 结语

世博展馆，本质上是展示各国文化内容的容器。塑造出其特有的场所精神无疑是每个参与其中的建筑师所追求的创作终极目标。两千多年前的老子对空间精粹就有过经典的论述:"埏埴以为器,当其无,有器之用;凿户牖以为室,当其无,有室之用"。然而空间在本质上具有社会性,同时社会也具有空间性,建筑师应通过规划和设计社会空间来服务于社会。因此可以说,从某种程度上,建筑已脱离了建筑师个人的美学范畴。世博会作为人类艺术和技术成果集中的大规模展示,其风向标作用是显而易见的,当然建筑包括在其中。作为全程参与其中的一份子,我们真心期待着这届世博会能为中国建筑师的创作带来清晰正确方向的推动,而不是相反!

本文引改自《建筑学报》2010(05)

西安咸阳国际机场新塔台及附属建筑 · Xi'an Xianyang International Airport New Tower and Annex Buildings

设计 Design 2009 · 竣工 Completion 2012

地点：陕西，西安 · 用地面积：2.53 公顷 · 建筑面积：5565 平方米
Location: Xi'an, Shaanxi · Site Area: 2.53ha · Floor Area: 5,565m²

合作设计师：刘嘉嘉、徐贵清、邓涛、郭珍珍、王凯、车爱明
Co-designer: Liu Jiajia, Xu Guiqing, Deng Tao, Guo Zhenzhen, Wang Kai, Che Aiming

西安咸阳国际机场新塔台及附属建筑是西安咸阳国际机场二期空管工程的配套工程。方案秉承整体性、功能性、经济性和先进性原则，注重彰显古城西安的传统内涵，设计在满足空中航管功能要求的前提下，将这一极具现代化功能的建筑赋予更多的文化和精神内涵。塔台建筑造型取自西安大雁塔、小雁塔的层叠、收分渐变的手法，形体玲珑秀丽、轮廓舒畅、比例优美；密檐式的百叶与玻璃幕墙的结合加深了建筑中的人文气息。塔身由实至虚，明室与检修环部位破茧而出，其造型语言结合现代机场导航塔台的功能特点来塑造，显得顺理成章；裙房的造型采用横向舒展的流线设计，肌理由密至疏，强化了塔台的高耸。统一的建筑手法与多方位的细部处理，恰如其分地呼应了机场高效、流畅的风格诉求，使整个建筑简约而不简单，精致优雅而不琐碎。

The new air traffic control tower of Xianyang International Airport and its affiliated building in Xi'an is a supporting project for the air traffic control project of Xi'an Xianyang International Airport Phase II. The design scheme adopts the principles of integrity, functionality, economy and advancement, paying attention on manifesting the traditional implication of Ancient Xi'an. On the precondition of meeting the requirements of flight navigation management function, the design endows more cultural and spiritual connotations to this building with extremely modern function. The architectural form of the control tower derives from the cascade and cone shape of the Big Wild Goose Pagoda and Small Wild Goose Pagoda; and the tower is exquisite and elegant, with a comfortable profile and delightful proportion. The combination of multi-eave louver and glass curtain wall enhances the humanistic air in the structure. With the tower body changing from the solid to the void and the component of the tower cab and inspection ring breaking out of the cocoon, it is only natural to combine the functional features of a navigation control tower for a modern airport to create the shape of the control tower. Horizontally extended streamline design is adopted for the shape of the podium, with the texture changing from dense to sparse, so as to highlight the loftiness of the control tower. Unified architectural techniques as well as the multi-dimensional detail treatment do justice to respond to the appeal of an airport for efficiency and swiftness; as a result, the whole structure looks simple but not simplistic, elegant yet uncluttered.

生长与蔓延　蔓·设计实践 I

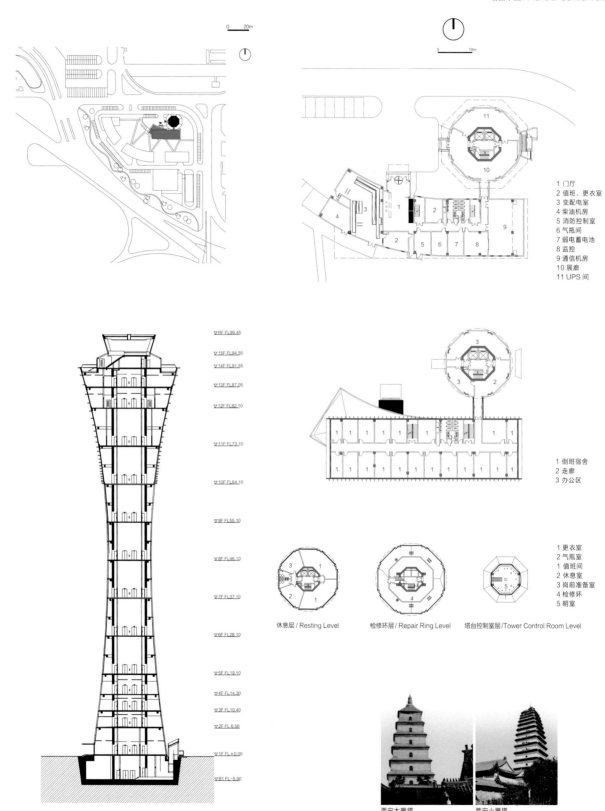

总平面 / General Layout
首层平面 / First Floor Plan
二层平面 / Second Floor Plan
塔台平面 / Plans of Control Tower

剖面 / Section
设计意向 / Intentions of Design

长沙中电软件园总部大楼 · The Headquarters Building of Changsha CEC Software Park

设计 Design 2009 · 竣工 Completion 2013
地点：湖南，长沙 · 用地面积：2.78 公顷 · 建筑面积：5.84 万平方米
Location: Changsha, Hunan · Site Area: 2.78ha · Floor Area: 58,400m²

合作设计师：权薇、朱谞、孙成伟、王少雷、李华、邓涛、车爱明、缪卓卫
Co-designer: Quan Wei, Zhu Xu, Sun Chengwei, Wang Shaolei, Li Hua, Deng Tao, Che Aiming, Miao Zhuowei

长沙中电软件园位于长沙麓谷国家高新技术产业开发区园内。其中总部大楼位于城市干道交叉口的西南角，西侧面向规划园区的中心景观湖面，自然环境优越宜人。总部大楼作为软件园内首幢标志性建筑，对整个园区的启动和发展起着至关重要的作用。设计因地制宜，在面向不同城市要素的界面做不同的表情处理，在两个方向、空间和形态上营造建筑与周边环境的和谐关系。将高耸的主楼置于面向城市主干道的一侧，挺拔端庄而富有仪式感，面向园区景观湖的一侧用半圆环形裙房拥抱中心湖景。为使80m限高下的主楼呈现出更具向上的张力，在其顶部和底部采用了倾斜10°的处理手法；为适应南方炎热的气候特点，立面选择内凹式开窗组合的竖向肌理，配以土黄色陶土面砖，进一步加强了建筑纵向的力量感。临湖一侧弧线形的职工餐厅在造型上采用外廊格栅式处理，突出建筑与中心湖景的渗透和融合，在遮阳的同时使建筑更具层次感。建筑整体注重传达湖湘文化大方、朴实和求变创新的精神气质。

Changsha CEC Software Park is located in Changsha Lugu National Hi-Tech Industry Development Zone.The headquarters building is on the southwest corner of the intersection of arterial streets, with the west side of the building facing the central waterscape lake in the planned park; the natural environment is attractive and delightful.As the first icon in the Software Park, the headquarters building plays a vital role for the startup and development of the whole Park.In the light of the specific conditions the design adopts different facial expressions for the interfaces facing different urban elements in a bid to create harmonious relations between buildings and their ambient environment in terms of space and shape in two directions.The soaring main building on the side facing the urban artery looks lofty and dignified with a sense of rite; the side facing the waterscape lake in the park embraces the central lakeview with its semi-circular podium.In order to make the main building with 80-meter limit appear to be more towering a 10-degree inclination is applied in both the bottom and top of the building.The vertical texture created by the indented windowing in consideration of the sweltering climate in south China, combined with the earthly yellow clay tiles, reinforces the sense of vertical strength of the building.The grating veranda outside the arc-shaped staff cafeteria on the side facing the lake highlights the infiltration and fusion between the building and the central lakeview, giving the building an extra layer while providing sunshade.The entirety of the headquarters building lays emphasis on convey the ethos of generosity, simplicity and innovation.

生长与蔓延　蔓·设计实践 I

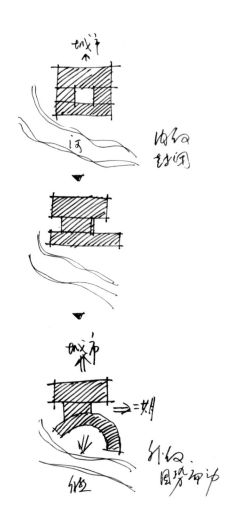

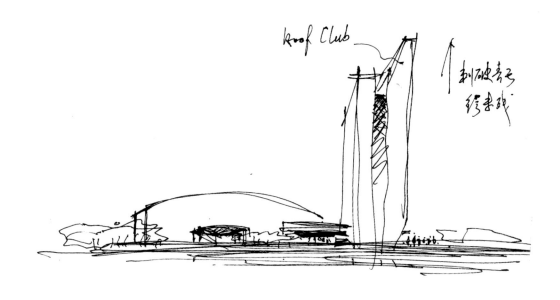

总平面 / General Layout
设计草图 / Sketch

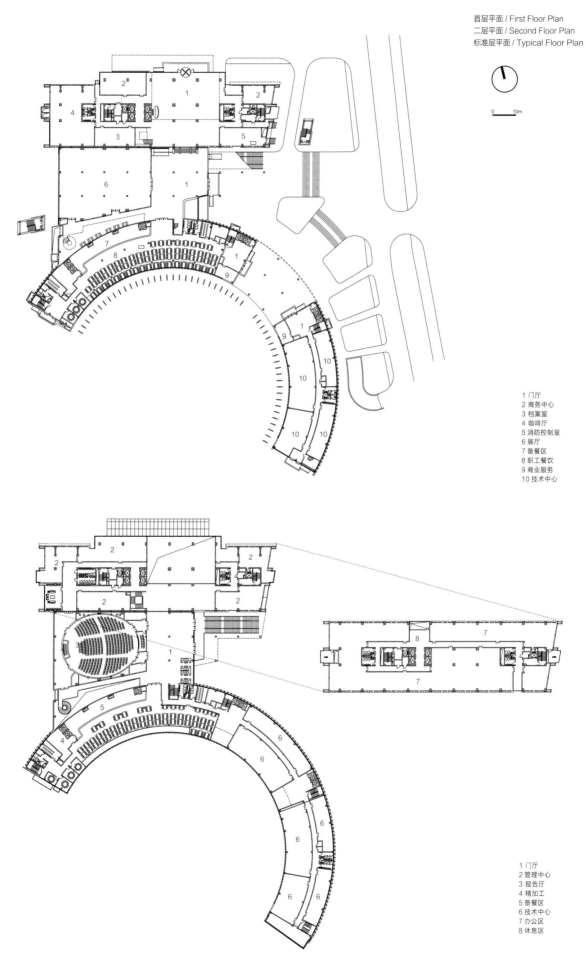

生长与蔓延　蔓·设计实践 I

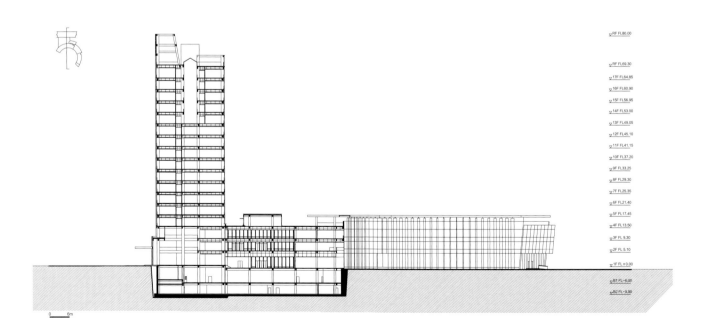

剖面 / Section

长沙中电软件园总部大楼

固安规划展馆 · Gu 'An Urban Planning Exhibition Hall

设计 Design 2010 · 竣工 Completion 2014

地点：河北，固安 · 用地面积：2.96 公顷 · 建筑面积：1.01 万平方米
Location: Gu 'an, Hebei · Site Area: 2.96ha · Floor Area: 10,100m²

合作设计师：权薇、孙成伟、邓涛、李华、缪卓卫、王凯、车爱明
Co-designer: Quan Wei, Sun Chengwei, Deng Tao, Li Hua, Miao Zhuowei, Wang Kai, Che Aiming

建筑以"城市魔方"的概念，将展馆体量用 3m 宽的网状空间划分为 16 个模块，然后根据城市展馆的功能需要进行平面和立体上的编织、组合，生成不同尺度、不同形式的使用功能空间，呈网状的线性空间既有导引、分隔作用，也是将自然光带入建筑的切入口。我们将其中的模块抽出，将自然庭院植入，有效地解决了展馆的通风和采光问题，为内部空间带来活力。建筑形象处理延续平面的构成关系，强调实与虚、进与退、厚重与透明的对比，加上灰色系中富有韵律的深浅搭配，使展馆既有远观时清晰的构成感，又有近看时细腻和丰富的质感。景观系统为突出建筑的立体感并考虑导览需要，将薄水面环绕四周，通过桥形通道导入透明的主入口空间。建筑用以大化小、化整为零的设计手法将自身与周围绿林和水景有机地融合在一起。

Utilizing the concept of "urban Rubik's Cube", the exhibition space is divided into 16 modules of 3-meter wide lattice space.Then weave and combine the blocks in both planar and spatial dimensions according to the functional requirements of the urban exhibition center to generate functional areas of different scales and forms.The lattice-like linear space serves not only as guides and partitions but also as entrances for natural light into the building.Modules within this framework are extracted to incorporate natural courtyards, effectively addressing ventilation and lighting issues within the exhibition hall, thereby infusing vitality into the interior space. The architectural imagery extends the compositional relationships of the plan, emphasizing the contrast between solid and void, advancing and receding, heaviness and transparency.Coupled with rhythmic variations in the grayscale palette, the exhibition hall possesses both a clear sense of composition when viewed from afar and a delicate, rich texture when observed up close.The landscape system, aiming to accentuate the building's three-dimensionality and considering navigation needs, surrounds the perimeter with thin water bodies, directing visitors into the transparent main entrance space via bridge-like passages.Through design techniques of maximizing space utilization and breaking down the whole into parts, the architecture seamlessly integrates with the surrounding greenery and water features.

生长与蔓延 蔓·设计实践 I

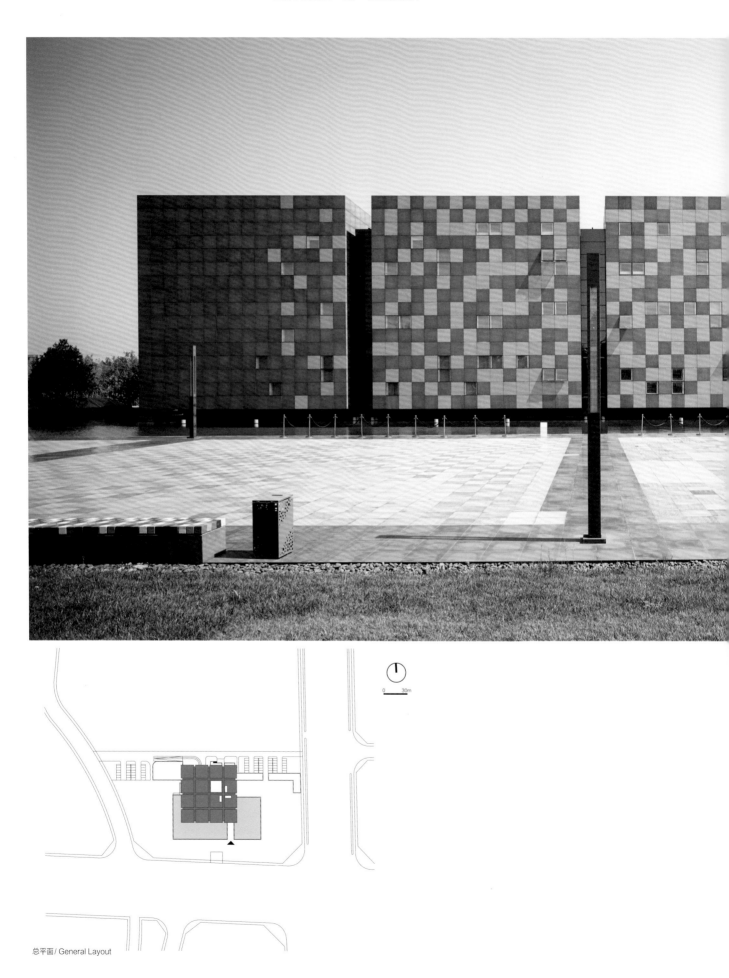

总平面 / General Layout

固安规划展馆

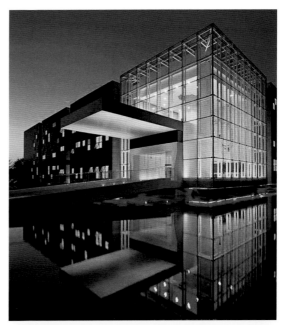

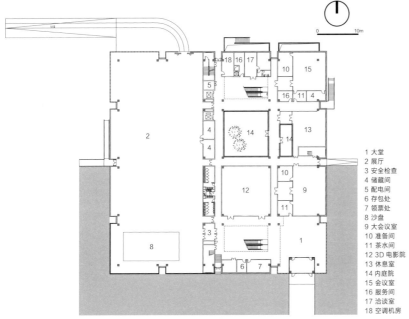

首层平面 / First Floor Plan
二层平面 / Second Floor Plan

1 大堂
2 展厅
3 安全检查
4 储藏间
5 配电间
6 存包处
7 领票处
8 沙盘
9 大会议室
10 准备间
11 茶水间
12 3D 电影院
13 休息室
14 内庭院
15 会议室
16 服务间
17 洽谈室
18 空调机房

1 洽谈室
2 服务间
3 多功能厅
4 大报告厅
5 开敞办公
6 接待室
7 秘书办公室
8 办公室
9 经理办公室
10 空调机房
11 配电间
12 小会议室
13 上空

北京泰德制药股份有限公司研发中心扩建 · R&D Center Expansion Project of Beijing Tide Pharmaceutical Co., Ltd.

设计 Design 2010 · 竣工 Completion 2014

地点：北京 · 用地面积：3.24 公顷 · 建筑面积：4.31 万平方米
Location: Beijing · Site Area: 3.24 ha · Floor Area: 43,100m²

合作设计师：王然、邓涛、陈珑、孙成伟、马光磊、李娜、尚晓松、扈秀英、梁晓、王利、康有财、刘澈、车爱明、王凯
Co-designer: Wang Ran, Deng Tao, Chen Long, Sun Chengwei, Ma Guanglei, Li Na, Shang Xiaosong, Hu Xiuying, Liang Xiao, Wang Li, Kang Youcai, Liu Che, Che Aiming, Wang Kai

　　该项目为一期工程的扩建工程，与厂区内现有建筑和空间环境的有机融合以及维护厂区现有生态系统是设计过程中始终遵循的原则。本设计通过L形体量与一期工程形成围合庭院，职工中心及后勤服务中心作为连接体，将一、二期主体联系在一起。主体设计通过打造共享绿色中庭，将工艺和实验功能结合在一起，为研发人员严谨、理性的工作状态带来了视觉和精神上的放松；建筑材料和构造的推敲与选择，以及室内效果，均注重强化科研建筑设计上对逻辑的迷恋和对制作工艺精致性的追求。本项目以美国制药标准进行设计，内部有小试、中试、研发、仓储、设备等主要功能，且各功能层高、洁净、振动要求均不相同。利用通高中庭将层高不同的小试、中试空间隔开，且将仓储、设备等功能巧妙布局，使各功能既保证了独立性，同时又有机统一地结合在一起。

This project is an expansion of Phase I and is designed to organically integrate with the existing buildings and spatial environment within the factory area, which has been a guiding principle throughout the design process. This design forms an enclosed courtyard with an L-shaped volume in conjunction with Phase I of the project. The staff center and logistics service center act as connecting elements, linking the main bodies of the first and second phases together. The main design combines process and experimental functions by creating a shared green atrium, providing visual and spiritual relaxation to the rigorous and rational working environment of R&D personnel. The consideration and selection of building materials and structures, as well as the interior design, all focus on emphasizing the fascination with logic and the pursuit of exquisite processes in the design of scientific research buildings. This project is designed according to American pharmaceutical standards and includes main functions such as small-scale testing, pilot testing, research and development, warehousing, and equipment, each with varying floor heights, cleanliness requirements, and vibration specifications. The design uses a high courtyard to separate small and medium-sized trial spaces with different levels of height, and cleverly layout functions such as storage and equipment, ensuring the independence of each function while organically integrating them together.

生长与蔓延　蔓·设计实践 I

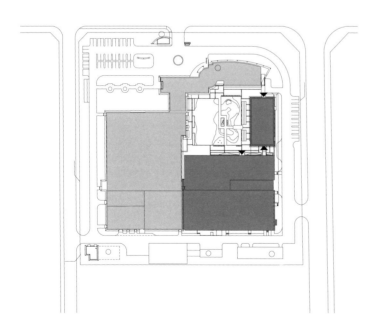

总平面 / General Layout

北京泰德制药股份有限公司研发中心扩建

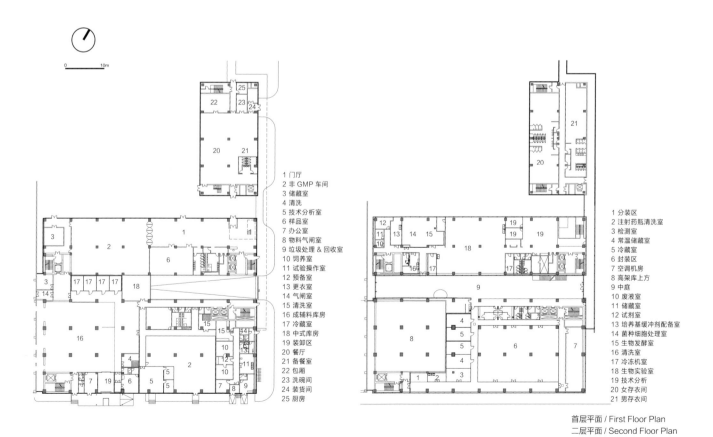

1 门厅
2 非 GMP 车间
3 储藏室
4 清洗
5 技术分析室
6 样品室
7 办公室
8 物料气闸室
9 垃圾处理 & 回收室
10 饲养室
11 试验操作室
12 预备室
13 更衣室
14 气闸室
15 清洗室
16 成辅料库房
17 冷藏室
18 中式库房
19 装卸区
20 餐厅
21 备餐室
22 包厢
23 洗碗间
24 装货间
25 厨房

1 分装区
2 注射药瓶清洗室
3 检测室
4 常温储藏室
5 冷藏室
6 封装区
7 空调机房
8 高架库上方
9 中庭
10 废液室
11 储藏室
12 试剂室
13 培养基缓冲剂配备室
14 菌种细胞处理室
15 生物发酵室
16 清洗室
17 冷冻机室
18 生物实验室
19 技术分析
20 女存衣间
21 男存衣间

首层平面 / First Floor Plan
二层平面 / Second Floor Plan

中国信达（合肥）灾备及后援基地·China Cinda (Hefei) Disaster Recovery and Back-Up Base

设计 Design 2012 · 竣工 Completion 2016

地点：安徽，合肥 · 用地面积：5.56 公顷 · 建筑面积：13.98 万平方米
Location：Hefei, Anhui · Site Area：5.56ha · Floor Area：139,800m²

合作设计师：蒋小华、李达、袁源、冯伟、陈珑、孙成伟、尹钰
Co-designer：Jiang Xiaohua, Li Da, Yuan Yuan, Feng Wei, Chen Long, Sun Chengwei, Yin Yu

本项目为中国信达公司位于合肥的灾备中心，是以数据储存、处理、维护、传输及综合分析等功能为核心的数据中心综合体。基地由数据、后援、研发和培训中心以景观中心庭院形式围合而成。四水归堂、藏风聚气的型制，既呼应了基地文脉也表达了对地域特色的关注。设计最大限度地整合利用城市景观资源的同时注重打造内部空间，使得整个建筑组群与自然交融在一起，以期为从事高强度的数据管理分析的研发和管理人员提供高质量的交流、聚会、相互激励和激发灵感的场所。数据中心被置于最安全和安静的基地东侧，朝南侧以灰空间手法向城市打开形成主入口；高层研发楼置于基地北侧以减少对庭院的遮挡；后勤中心处在西侧城市小型街边公园与中庭的围合中，以柔性裙房与U形体块设计增加与中庭环境的接触面积。面向城市的一面继续采用连续界面手法将建筑组群有机地整合在一起，而将基地核心——数据中心处理成具有地域特点的新徽派建筑风格，以体现文化传承和人文关怀；基地制高点——研发大楼向主干道方向张开的同时与远处的巢湖风景区作出呼应。建筑外立面材料选用温润柔和的陶土板搭配简洁大方的玻璃幕墙，形成天然与人工、传统与现代、感性与理性的对比，二者刚柔相济。

The China Cinda (Hefei) Disaster Recovery and Back-up Base is a multifunctional data center complex, offering a range of functions such as data storage, processing, and analysis, that harmonizes with its natural and cultural surroundings. The complex, structured around a central courtyard, features dedicated areas for data handling, support, research and development, and training. It aims to blend with nature and the urban landscape, providing a conducive environment for data management and analysis professionals to collaborate and innovate. The data center, positioned for optimal safety and quietness, faces south, integrating with the urban landscape. A high-rise research building on the north side and a logistics center with a flexible design on the west side frame the central courtyard, enhancing connectivity with nature. The complex adopts a Huizhou architectural style for the data center, embodying regional cultural values and humanistic considerations. The research building faces the main road and nearby Nest Lake Scenic Area, making a statement with its strategic location. The selection of exterior facade materials combines the use of warm and gentle terracotta panels with sleek and elegant glass curtain walls, creating a contrast between the natural and the artificial, the traditional and the modern, and the emotional and the rational. This combination achieves a harmonious balance between rigidity and softness.

生长与蔓延　蔓·设计实践 I

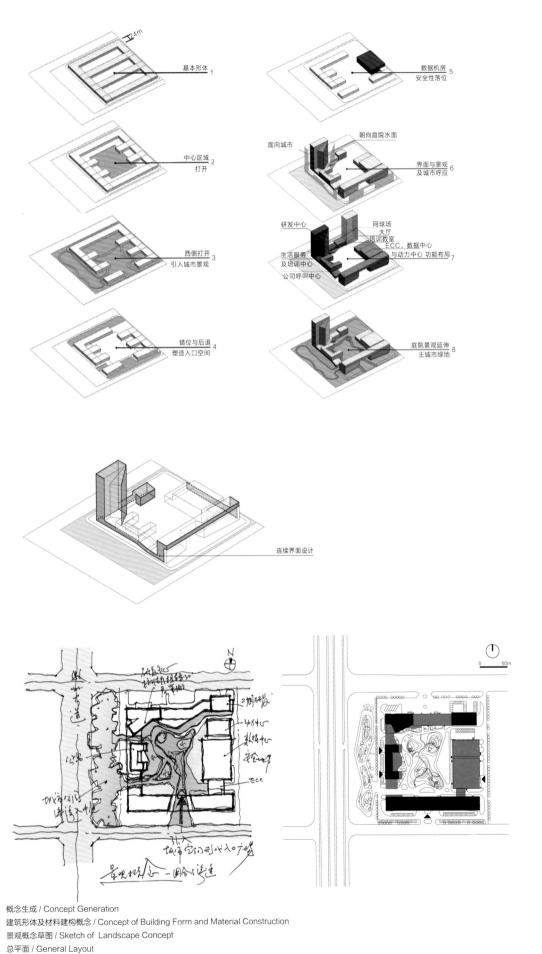

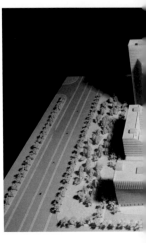

概念生成 / Concept Generation
建筑形体及材料建构概念 / Concept of Building Form and Material Construction
景观概念草图 / Sketch of Landscape Concept
总平面 / General Layout

中国信达（合肥）灾备及后援基地

模型 / Models

生长与蔓延　蔓·设计实践 I

首层平面 / First Floor Plan
三层平面 / Third Floor Plan
标准层平面 / Typical Floor Plan

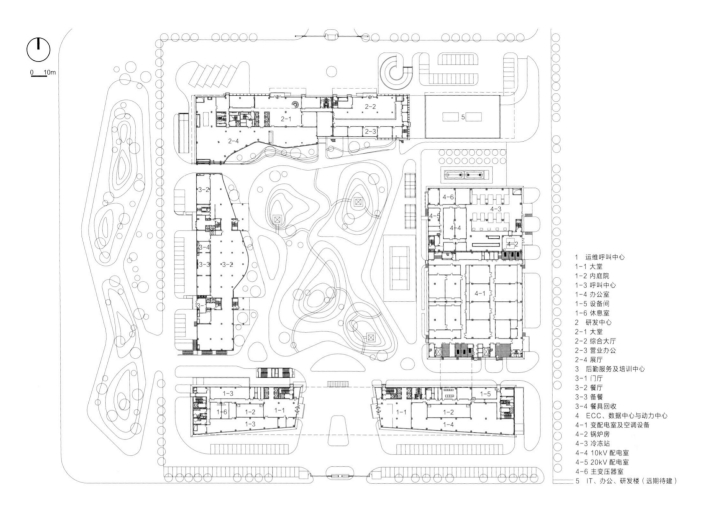

1　运维呼叫中心
1-1　大堂
1-2　内庭院
1-3　呼叫中心
1-4　办公室
1-5　设备间
1-6　休息室
2　研发中心
2-1　大堂
2-2　综合大厅
2-3　营业办公
2-4　展厅
3　后勤服务及培训中心
3-1　门厅
3-2　餐厅
3-3　备餐
3-4　餐具回收
4　ECC、数据中心与动力中心
4-1　变配电室及空调设备
4-2　锅炉房
4-3　冷冻站
4-4　10kV 配电室
4-5　20kV 配电室
4-6　主变压器室
5　IT、办公、研发楼（远期待建）

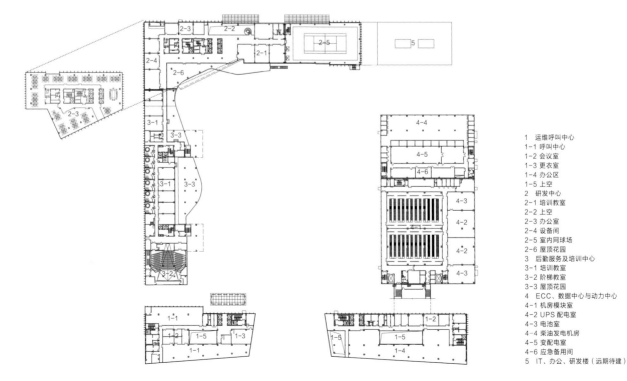

1　运维呼叫中心
1-1　呼叫中心
1-2　会议室
1-3　更衣室
1-4　办公区
1-5　上空
2　研发中心
2-1　培训教室
2-2　上空
2-3　办公室
2-4　设备间
2-5　室内球场
2-6　屋顶花园
3　后勤服务及培训中心
3-1　培训教室
3-2　阶梯教室
3-3　屋顶花园
4　ECC、数据中心与动力中心
4-1　机房模块室
4-2　UPS 配电室
4-3　电池室
4-4　柴油发电机房
4-5　变配电室
4-6　应急备用室
5　IT、办公、研发楼（远期待建）

中国信达（合肥）灾备及后援基地

生长与蔓延　蔓·设计实践 I

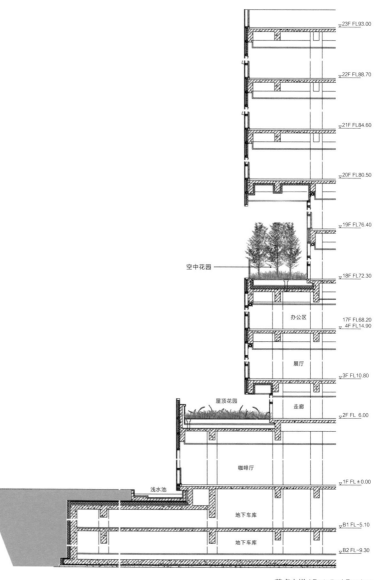

节点大样 / Detailed Design

庭·源——中国信达（合肥）灾备及后援中心

王振军　李　达　孙成伟

该项目为中国信达公司"二地三中心"灾备模式下的合肥中心建设项目，位于合肥城市发展轴线上的滨湖新区，徽州大道与杭州路交叉口，距离巢湖 3km。其主要功能包含数据中心、动力中心、总控中心（ECC）、运维呼叫中心、配套的研发中心及后勤服务培训中心。项目建设用地 5.56 公顷，总建筑面积 13.98 万 m^2（其中地上 10.36 万 m^2）。

对于数据灾备这一典型的 IT 建筑设计而言，我们面对的最根本的任务依然是"如何处理建筑与自然、人与自然的关系"这一本源问题。设计基于生态的有机建构理念，遵循"四水归堂"、返璞归真、道法自然、藏风聚气的传统哲学思想，最大限度地整合利用基地外部环境资源的同时注重打造内部景观环境，使自然、人与建筑之间形成极度和谐、交融共生并可持续发展的生态秩序。

该基地的数据、后援、研发和后勤中心以两个 L 形体块相扣有机地围合，构成幽静怡人的"景观合院空间"形式，成为整个建筑群的物理空间中心。其中数据机房基于其高安全性运维需求，置于受干扰最小的用地东侧；员工较集中的研发中心、后勤服务及培训中心则布置于用地的西北角和西侧，处在西侧城市小型街边公园与中庭的围合空间中。

中心庭院作为整个建筑机体系统中的"绿色心脏"，其主体设计运用生动的造园手法，营造富有层次的立体开敞景观，并集中布置疏密有致的树阵及水面。置身于内庭院中的人们可以体验到退台屋顶花园的层次感、架空连廊塑造的灰空间等多维度变化。

不仅如此，面向城市的一面继续采用连续界面手法将建筑组群有机地整合在一起，塑造了空间的整体氛围。这些具有音乐的韵律的形态，如建筑的开合、扭转、错落、架空等均充分考虑了城市文脉及周围的景观脉络，达到庭院景观与南部的城市空间、西侧的城市绿地空间之间的互通，形成自然环境的融合渗透和互动协调。

基地的核心——数据中心建筑处理成具有地域特点的新徽派建筑风格，以体现建筑的地域性文化传承和人文关怀；基地制高点——研发大楼向主干道方向张开的同时与远处的巢湖风景区作出呼应，站在研发大楼十八层的空中庭院眺望，巢湖美景尽收眼底。

建筑的外立面材料选用温润柔和的陶土板搭配简洁大方玻璃幕墙的方式，形成天然与人工、传统与现代、感性与理性的对比，二者刚柔并济；朝向内庭的一侧使用反射度低的玻璃幕墙材料，利用镜面反射原理形成庭院尺度无限深远扩大的视觉效果，将自然要素和人的活动进一步放大和夸张，强化了庭院的重要性和场所感。

本项目试图从生态学角度创作，以自然作为生发所有要素的源头，以庭院作为协调人、技术和建筑之间关系的中心，以期为从事高强度的数据管理分析的研发和管理人员提供高质量的交流、聚会、相互激励和激发灵感的场所。

本文引改自谷德设计网 2017 年 7 月

郑州新郑国际机场二期扩建空管工程塔台小区土建及配套工程 · Control Tower and Supporting buildings of Zhengzhou Xinzheng International Airport expansion Project phase II

设计 Design 2013 · 竣工 Completion 2016

地点：河南，郑州 · 用地面积：2.13公顷 · 建筑面积：1.32万平方米
Location：Zhengzhou, Henan · Site Area: 2.13ha · Floor Area: 13,200m²

合作设计师：邓涛、朱谓、李达、陈珑、孙成伟、李娜、车爱明、杨永斌、缪卓卫、蔡广会、郭未成
Co-designer: Deng Tao, Zhu Xu, Li Da, Chen Long, Sun Chengwei, Li Na, Che Aiming, Yang Yongbin, Miao Zhuowei, Cai Guanghui, Guo Weicheng

郑州新郑国际机场地处中原腹地，是我国最大的重要干线机场及空中交通枢纽之一。项目位于郑州新郑国际机场综合交通换乘中心（GTC）交通枢纽北侧，包含塔台、航管楼两部分。方案从空港宏观环境分析入手，结合地域文化元素，造型取自河南重大考古发现——"贾湖骨笛"（现为河南省博物馆的镇馆之宝）与商代青铜酒器"觚"的形象，以谦逊的弧形姿态伫立在机场轴线北侧，以烘托和强调机场主轴线及航站楼的主角地位，并与航站楼的流线型设计相得益彰，相辅相成，以提高机场总体环境的整体性和协调性。新郑塔台是我国首座最高的非线性清水混凝土建筑，高93.5m。在满足使用功能要求下，采用了国际领先的建筑型制、经济高效的结构选型及高集成化的施工工艺。项目在方案创作、施工图设计、设计配合施工等全阶段中应用BIM技术，保障了设计完成度和施工精度，并使后续使用、维护及管理更加高效。建成后的塔台以和谐的建筑形态融入城市空间，以谦逊之姿、欢迎之姿迎来送往，全面展示了郑州大气、庄重、古韵的城市特质，彰显出"中原崛起"的时代精神。

Zhengzhou Xinzheng International Airport is located in the hinterland of Central Plains, and it is one of China's biggest trunk line airports and air traffic hubs.This project is on the north of the Zhengzhou Xinzheng International Airport GTC, including a control tower and an air traffic control building.The design scheme analyses the project factors from macro-environment, combined with the regional cultural elements.The shape of the control tower originates the major archaeological discovery in Henan Province—"Jiahu bone flute"（it is currently the centerpiece of Henan Museum）, and the image of "Gu", a kind of bronze drinking vessel popular in the Shang dynasty (ca.1600 B.C.~ ca.1046 B.C.) .The control tower stands on the north side of the airport axis with its humble arc-shaped posture, so as to foster and emphasize the leading role position of the airport axis and terminal, benefitting a company with the streamlined design of the terminal; hence, the integrity and coherence of the overall environment of the airport is uplifted.The Xinzheng control tower is China's first and highest non-linear fair-faced concrete structure, with the apex of 93.5m.While the requirement of utility function is met, what is applied in the design and construction processes are internationally advanced architectural type, economically efficient structure selection, as well as highly integrated construction technology.BIM technique is applied throughout the design process, from conceptual design to construction documents design, and from the design coordination to construction, so that the design schedule and construction accuracy are guaranteed; and subsequent utilization, maintenance and management are more efficient.The as-built control tower blends in with the urban space with its harmonious architectural shape, greeting and bidding farewell with the humble and welcoming posture.It displays Zhengzhou's urban traits as being grand, solemn and archaic, manifesting the ethos of the time during the "rise of the Central Plains".

生长与蔓延　蔓·设计实践 I

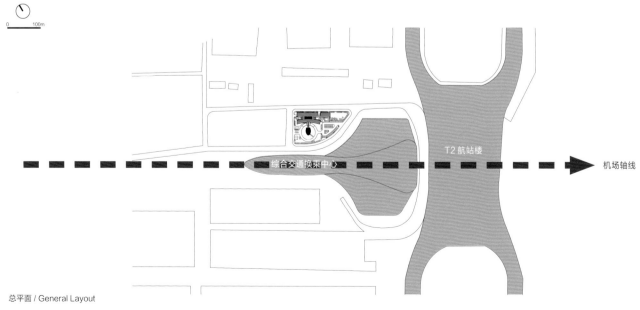

总平面 / General Layout

郑州新郑国际机场二期扩建空管工程塔台小区土建及配套工程

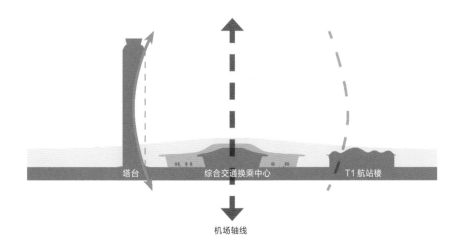

塔台　综合交通换乘中心　T1航站楼

机场轴线

轴线分析 / Axis Analysis

（1）城市空间的诠释——谦逊之姿、欢迎之姿

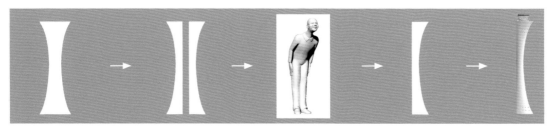

（2）塔造型的文化意象之一——"贾湖骨笛"

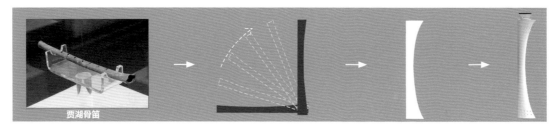

（3）塔造型的文化意象之二——"觚"

概念生成 / Concept Generation
模型 / Model

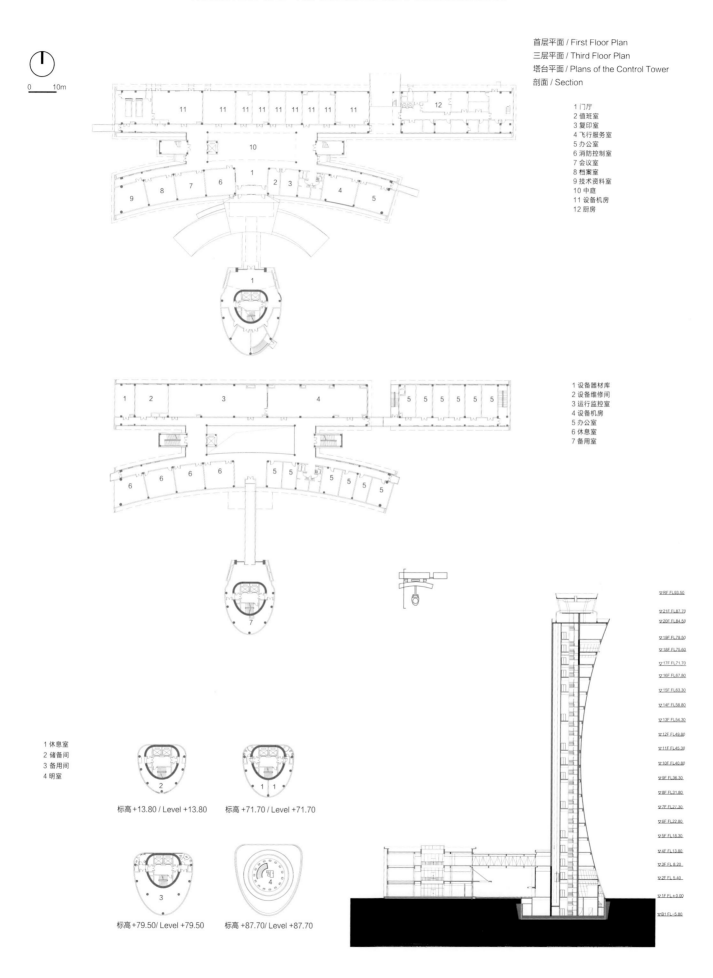

塔的故事——郑州新郑国际机场塔台及航管楼设计

王振军

项目概况

郑州新郑国际机场地处中原腹地，是我国重要干线机场及空中交通枢纽之一。本工程是郑州新郑国际机场的配套工程，现已与郑州新郑国际机场二期扩建工程同步投入使用。

新郑国际机场新塔台及航管楼项目位于郑州新郑国际机场综合交通换乘中心（GTC）交通枢纽北侧，用地面积为21366.3m^2，包含塔台、航管楼两部分（其中航管楼包括办公和生活配套，两部分通过连廊连接）。航管楼总建筑面积6205m^2，新塔台最高点高93.5m。

提起机场，人们马上想到的是"城市的门户""远行的起点和终点""迎来送往""悲欢离合"等。毫无疑问，城市的机场已是现代城市生活越来越重要的节点，而导航塔台作为掌控机场整体秩序的核心"指挥官"，以其修长出众的高度往往会成为机场的标志物而引起公众关注。因而，其设计无论从形态、空间、功能，还是对城市的情感意义上，都值得推敲斟酌。

设计概念

整个方案设计秉承整体性、功能性、经济性和先进性原则，注重从宏观环境考量项目要素，并结合地域文化元素，体现对"中原文化"文脉传承的同时，以其非线性、双曲面的独特造型诠释了郑州古都别具一格的创新追求，充分展现了"中原文化"的传统内涵与新空港时代精神的巧妙融合。

设计在满足塔台空中管制、导航的功能要求的前提下，将这一极具现代化功能的建筑赋予了更多的文化含义和精神内涵：

1. 城市空间的诠释——谦逊之姿、欢迎之姿

塔台显要地矗立在机场入口中轴线的北侧，是轴线一旁的最佳配角，与轴线上T1、T2航站楼和综合交通换乘中心（GTC），形成交相辉映的整体构图。相对于垂直的立面造型，塔台面向中轴的一侧采用月牙弧度造型，更好地与机场主体建筑融合呼应，以更谦逊的姿态烘托和强调轴线，凸显交通枢纽与航站楼形成的机场主角地位，以更和谐的建筑形态融入城市空间，以谦逊之姿、欢迎之姿迎来送往。

2. 非线性造型与航站楼呼应——交相辉映，融为一体

塔台建筑采用非线性的曲线造型，与航站楼的流线型设计相得益彰，相辅相成，提高了新郑机场总体环境的整体性和协调性。

3. 塔台造型的文化意向

（1）文化意向之一——贾湖骨笛

新塔台的造型取自河南省博物馆的镇馆之宝——贾湖骨笛的弧线外形。原因有三：其一，"骨笛"为河南本地出土，历史悠久，暗喻建筑为凝固的音乐；其二，"骨笛"极似塔形，修长优美；其三，弧线形塔台与柔美的航站楼在外形上取得呼应。整个塔台简洁大方，大巧若拙，平和而不平庸。

（2）文化意向之二——觚

新塔台北侧呈曲线型，取自商代青铜酒器"觚"的形态。原因有三：其一，"觚"具有端庄典雅，大气尊贵的气质，且其酒具的用途贴切地契合了机场分别与欢聚时觥筹交错的城市空间特性；其二，青铜文化象征和代表着河南鼎盛时期独特的古典文化底蕴；其三，"觚"的意向也寓意了河南鼎盛而独特的姿态，沧桑但厚积薄发的屹世雄心。

材料特点

新郑塔台是我国首座最高的非线性清水混凝土建筑，采用清水混凝土现场浇筑，原因如下：

（1）清水混凝土具有朴实无华、自然沉稳的外观韵味和与生俱来的厚重、清雅的气质，这种高贵的朴素感非常贴切地诠释了新塔台的设计理念；

（2）清水混凝土一次成型，与其他装饰材料相比，更耐久且无需维护；

（3）清水混凝土是名副其实的绿色建材，混凝土结构不需要装饰，舍去了涂料、饰面等化工产品。结构一次成形，不剔凿修补、不抹灰，减少了大量建筑垃圾，免去了交通建筑外墙维护之困。考虑到建成效果，在多种模板分割形式上进行优化推敲，对塔台形体合理分格优化，最终在满足模板制作需求与保证建成效果的同时，减少了75%的模板数量。

技术应用

项目从方案到施工建成实现全过程设计，始终秉承工作室"蔓·设计"理念，施工精细质量良好，重点应用以下技术：

1. 在满足国内机场指挥塔使用功能要求下，采用了国际领先的建筑型制：

（1）高度集成化的小明室

人均使用面积小于国内平均值（国内现有明室均大于人均 8.4m^2），增加了明室内工作人员观测的便利性；

（2）明室检修环下置

打破了国内传统的检修环与明室平层布置的模式，既保证了明室导航设备的检修维护方便、快捷，也能使导航员席位尽可能贴近明室外窗玻璃，从而最大限度地保证导航员视线无遮挡，也避免了导航与设备维护间的相互干扰。

2. 结构选型经济高效

指挥明室设计采用了目前国内塔台明室最少的四根柱支撑，保证了明室航管员视线的通透。指挥室采用四根圆形开口钢柱，支撑了现浇刚性屋盖结构，将管线置入圆形钢柱之中。整体方案具有极好的刚性，满足地震作用下层间位移要求。

3. 高集成化的工艺设计

指挥间建筑面积小，布局紧凑合理。集成化程度高，使管制员的观察下视角较大（大于或等于 60°），可以在座位上很方便地观察机场跑道、滑行道、站坪上飞机的情况，大大减轻了管制员的劳动强度。塔台指挥明室全部采用国际最先进的管制设备，设备放在塔台指挥明室地下的设备维护夹层中，机务人员可方便地进行设备维护，避免了管制和设备维护的相互干扰。

4.BIM 技术的全过程应用

设计在方案创作、施工图设计、设计配合施工等全阶段中应用 BIM 技术，不仅解决了曲线造型在形体推敲、施工图定位和现场施工等环节中产生的问题，还确保各种公用设备管线和工程综合管线的精确分布，保障了项目的完成度，从而实现精细化设计、高精度施工、成本缩减以及高效运维管理。

项目自建成启用以来，大大提高了城市空管保障能力，较好地满足了航班量快速增长的需求，为飞行的安全、顺畅提供了强有力的保障。与此同时，塔台以其高耸优美、富有深意而别具一格的"指挥官"形象，已成为航空港区乃至郑州市的城市新标志。

本文引改自 WZJSTUDIO 公众号 2017 年 8 月

中央美术学院燕郊校区教学楼·Teaching Building of Central Academy of Fine Arts Yanjiao Campus

设计 Design 2013·竣工 Completion 2017

地点：河北，三河·用地面积：13.69公顷·建筑面积：3.43万平方米
Location：Sanhe, Hebei·Site Area：13.69ha·Floor Area：34,300m²

合作设计师：王铁华、陈珑、孙成伟、邓涛、李达、夏璐、徐彤、赵翀玺
Co-designer：Wang Tiehua, Chen Long, Sun Chengwei, Deng Tao, Li Da, Xia Lu, Xu Tong, Zhao Chongxi

　　本项目是中央美术学院燕郊校区的标志性主体建筑，位于校区的北侧主入口广场，建筑将矩形用地做减法处理，面向入口切出弧形用地，形成西广场，东侧面临校区南北主干道，做U形开放式庭院。建筑主立面局部向外翻折，象征着翻起的书页，丰富的体量配以玻璃幕墙及石材的立面材料组合，展现了建筑形体的雕塑感，使自然要素充分融入。功能布置在充分满足美术教学需求的前提下，强调动与静、教与学，以及展陈临时与长期之间转换的灵活性。中庭空间作为校园和教学空间的中介，予以重点打造，强调开放和变化，并将结构构件参与到空间构成中。不同尺度和开放度的中庭，栏板层层后退，直通屋面，柔美的曲线和强烈的秩序感，充分展现了美术教学楼的艺术性。东部跨度达42m的高空展厅，用桁架结构完美呈现，也展示出了校园建筑的科技感和未来感。

The project serves as the flagship structure at the Yanjiao Campus of the Central Academy of Fine Arts, situated at the northern entrance plaza of the campus.The building adopts a subtractive approach to the rectangular site. At the entrance, a curved plot of land was cut to form the West Square, with the east side facing the north-south main road of the campus, and a U-shaped open courtyard was designed. The main facade of the building is partially folded outward, symbolizing the flipped pages of the book. The richly sculpted volumes, combined with a facade material composition of glass curtain walls and stone, imbue the architecture with a sense of sculpture, facilitating the integration of natural elements. On the premise of fully meeting the needs of art teaching, the functional layout emphasizes the flexibility of transition between movement and stillness, teaching and learning, and the transition between temporary and long-term exhibition. The central atrium, as a mediator between campus and instructional spaces, receives particular attention, emphasizing openness and variability while involving structural elements in spatial composition.The atriums, varying in scale and openness, cascade backward with layers of balustrades, extending to the roof, showcasing the artistic quality of the fine arts teaching building.The high-span exhibition hall in the east, spanning 42 m is constructed with a truss structure, highlighting the technological and futuristic aspects of campus architecture.

生长与蔓延　蔓·设计实践 I

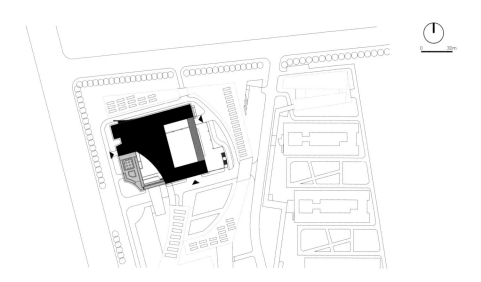

总平面 / General Layout

中央美术学院燕郊校区教学楼

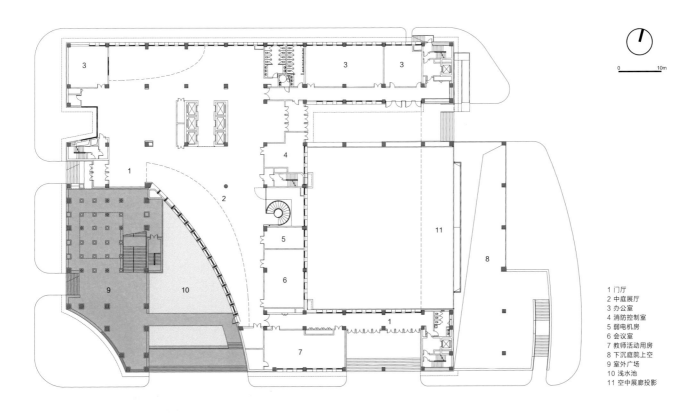

1 门厅
2 中庭展厅
3 办公室
4 消防控制室
5 弱电机房
6 会议室
7 教师活动用房
8 下沉庭院上空
9 室外广场
10 浅水池
11 空中展廊投影

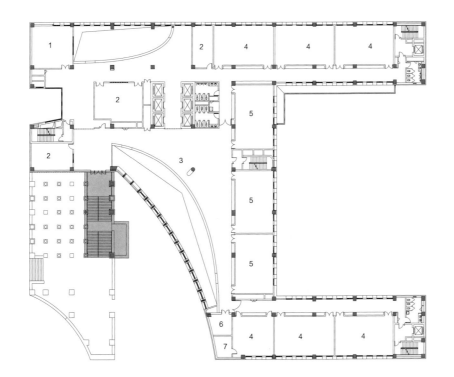

1 教研室
2 办公室
3 展廊
4 文化课教室
5 语音教室
6 强电间
7 储藏间

首层平面 / First Floor Plan
二层平面 / Second Floor Plan

燕郊世界华人收藏博物馆·Yanjiao Collection Museum of Worldwide Chinese

设计 Design 2013 · 竣工 Completion 2018

地点：河北，燕郊·用地面积：4.67 公顷·建筑面积：4.93 万平方米
Location：Yanjiao, Hebei · Site Area：4.67ha · Floor Area：49,300m²

合作设计师：王铁华、陈珑、孙成伟、徐彤、邓涛、马光磊
Co-designer：Wang Tiehua, Chen Long, Sun Chengwei, Xu Tong, Deng Tao, Ma Guanglei

本项目由博物馆主馆、副馆和收藏交流接待中心组成，主副馆通过室外连廊相连接，使二者在使用上能分能和，灵活多变。三座建筑用不同维度和尺度的庭院串联在一起，主馆入口设在二层，由大台阶引导而上。展馆围绕具有雕塑感的中庭布置，展览流线由上及下至地下收藏研究室；副馆以高大空间为主以适应展品拍卖功能，交流中心为会所和酒店功能。在材料运用上，建筑外立面以锈钢板为主要材料，配以象形文字的镂空处理，彰显建筑的古朴厚重。外立面条形窗与屋面的线性采光窗贯通，实现了形式和功能的统一。

The project comprises the main museum building, the auxiliary building, and the Collection Exchange and Reception Center. The main and auxiliary buildings are connected by an outdoor corridor, enabling them to function independently and flexibly. The three buildings are linked by courtyards of varying dimensions and scales. The entrance to the main building is situated on the second floor, guided via a grand staircase. The exhibition halls are arranged around a central atrium with sculptural elements, with the exhibition circulation descending from top to bottom towards the underground collection research rooms. The auxiliary building primarily features lofty spaces to accommodate exhibit auctions, while the exchange center serves as a club and hotel. In terms of materials, the exterior facade predominantly employs stainless steel panels, complemented by relief treatments inspired by hieroglyphs, highlighting the building's antique and substantial characteristics. The linear skylights on the facade and roof are seamlessly interconnected, achieving a harmonious fusion of form and function.

生长与蔓延 蔓·设计实践 I

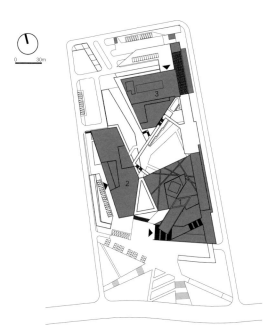

1 博物馆主馆
2 博物馆副馆
3 收藏交流接待中心

总平面 / General Layout

燕郊世界华人收藏博物馆

首层平面 / First Floor Plan
二层平面 / Second Floor Plan

1 门厅
2 中庭
3 石碑馆
4 陶瓷馆
5 青铜馆
6 阅览室
7 临时展厅
8 报告厅
9 前厅
10 放映室
11 后台
12 馆长办公室
13 馆长助理办公室
14 贵宾接待室
15 空调机房
16 修缮工作室
17 物理化学实验室
18 熏蒸室
19 空调机房
20 配电间
21 电气值班室
22 钢瓶间
23 藏品库房
24 藏品鉴定
25 礼品店

1 大堂
2 入口平台
3 室外雕塑平台
4 绿地
5 戊类物品储藏间
6 熏蒸室
7 修缮工作室
8 儿童乐园
9 影视厅
10 放映室
11 物品寄存
12 休闲茶吧
13 宋代瓷器展厅
14 龙泉窑展厅
15 电气用房

海盐杭州湾智能制造创新中心·Haiyan Intelligent Manufacturing Innovation Center

设计 Design 2014·竣工 Completion 2018

地点：浙江，海盐·用地面积：15.16 公顷·建筑面积：26.95 万平方米
Location：Haiyan, Zhejiang·Site Area：15.16ha·Floor Area：269,500m²

合作设计师：陈珑、李达、夏璐、孙成伟、徐彤、邓涛、朱谞、王沛、车爱明、王凯
Co-designer: Chen Long, Li Da, Xia Lu, Sun Chengwei, Xu Tong, Deng Tao, Zhu Xu, Wang Pei, Che Aiming, Wang Kai

　　基地位于国家中欧合作试点的海盐县，临近滨海公园，这可能是中国距大海最近的园区，距离杭州湾不到800m。规划以交融为概念，希望城市、海洋、人、科技与自然在此交织，最终交融在自然的脉络中。建筑高度和体量由北向南朝大海方向逐渐降低和缩小；环形主干道加组团支路的交通系统为中心景观湖实现纯步行化提供了保障；单体建筑将交通及辅助空间作模块化处理，与主研发楼体量组合在一起，以期为园区建设阶段的产业化提供可能。作为园区中心和制高点的总部大楼，以V字形平面面向大海方向，试图获得最大化的观海面，同时使北侧自然形成欢迎之态的主入口广场空间；而空间和形式的处理则通过左右两个体量的错动、穿插、咬合等手法，以追求建筑的动态平衡和左右两个体量关系的相互协调，同时寄予"合作、信心"等美好的期许。总部大楼成为园区的标志性建筑，更是观享海景的最佳位置。

This might be the science park that is closest to the sea in China as the site is less than 800 m from Hangzhou Bay. The science park is situated in Haiyan, where the National Sino-Europe Cooperation Pilot Program is carried out, and the coastal park is nearby. With the concept of blending as the guiding principle, the planning aims to interweave the city with sea, people, science and nature, with all elements finally interlacing into the natural context. The building height and mass gradually descends from the north to the south towards the direction of sea. The traffic system consisting of the circular main road plus access roads to building clusters ensures that only sidewalks are available along the central waterscape pond. Modularization of traffic and supporting space is done for individual buildings. They are combined with the main building for R&D. With all these buildings in place, it is intended to ensure the industrialization for the park even during the construction phase. As the center and commanding height of the park, the headquarters building faces the sea with the V-shaped sector in an attempt to have a maximum view of the sea; meanwhile, a greeting square space is formed naturally at the main entrance on the north side. For the design of space and shape, the application of such approaches as dislocation, interspersing and snap-in of the two masses on the left and right side seeks a dynamic balance of the buildings and inter-coordination between them. The relationship of the two interactive buildings also implies the expectation of "cooperation, faith and confidence". The headquarters building is the icon of the park and also the optimal place to enjoy the seascape.

生长与蔓延 蔓·设计实践 I

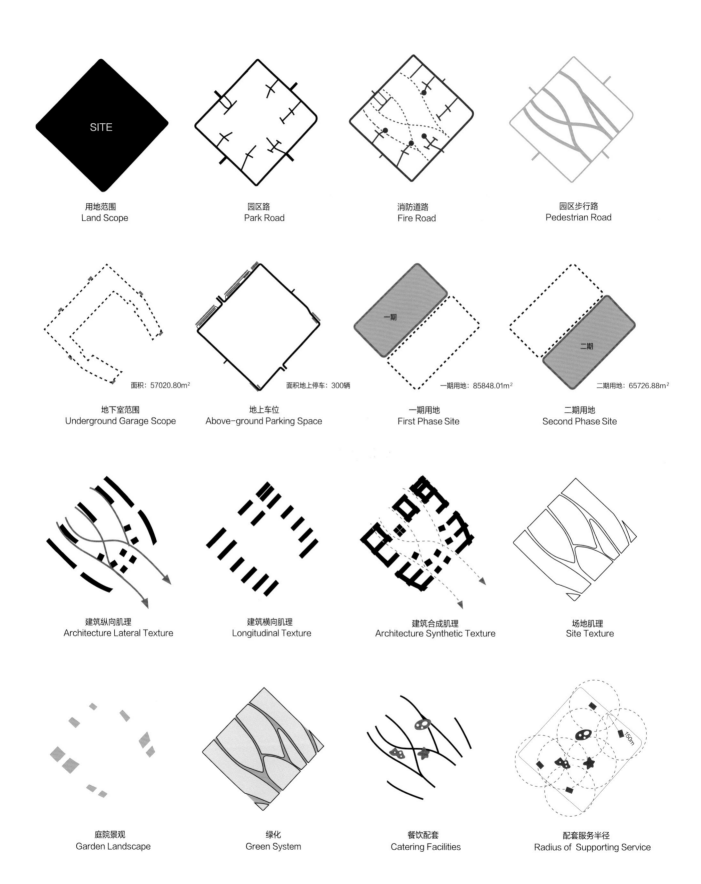

概念生成 / Concept Generation

海盐杭州湾智能制造创新中心

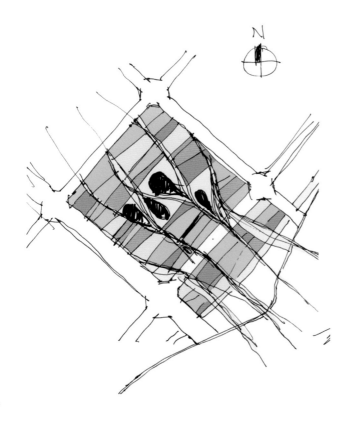

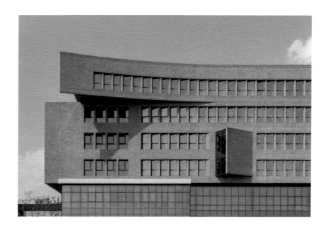

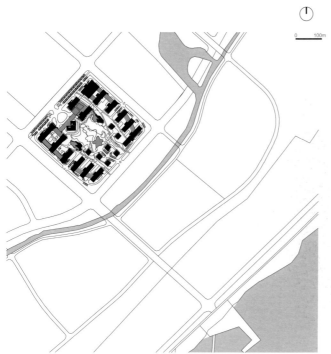

设计草图 / Sketch
总平面 / General Layout

163

海盐杭州湾智能制造创新中心

生长与蔓延　蔓·设计实践 I

首层平面 / First Floor Plan
二层平面 / Second Floor Plan

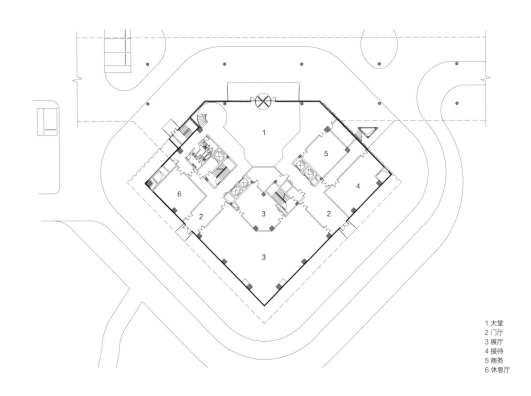

1 大堂
2 门厅
3 展厅
4 接待
5 商务
6 休息厅

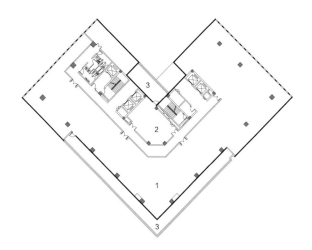

1 研发办公区
2 会议室
3 屋顶花园

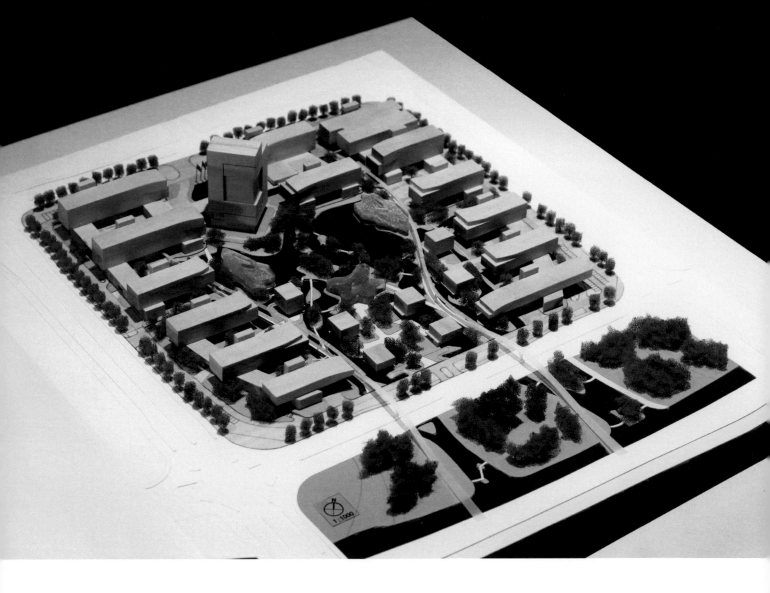

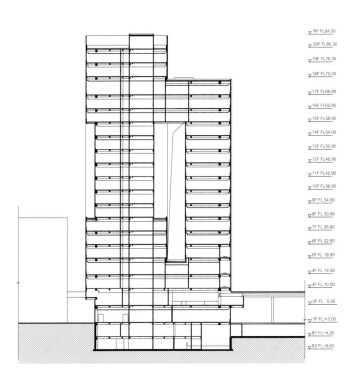

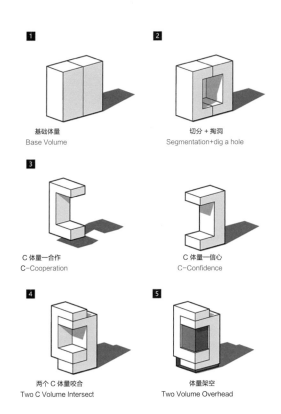

模型 / Model
剖面 / Section
总部大楼体量生成 / Volume Generation

1 基础体量 / Base Volume
2 切分 + 掏洞 / Segmentation+dig a hole
3 C 体量—合作 / C-Cooperation
　C 体量—信心 / C-Confidence
4 两个 C 体量咬合 / Two C Volume Intersect
5 体量架空 / Two Volume Overhead

生长与蔓延　蔓·设计实践 I

海盐杭州湾智能制造创新中心

海盐杭州湾智能制造创新中心

生长与蔓延　蔓·设计实践 I

餐厅首层平面 / First Floor Plan
餐厅二层平面 / Second Floor Plan
餐厅剖面 / Section

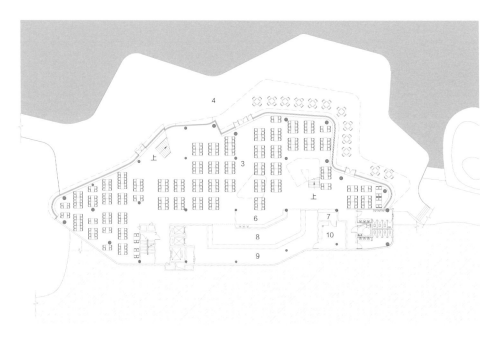

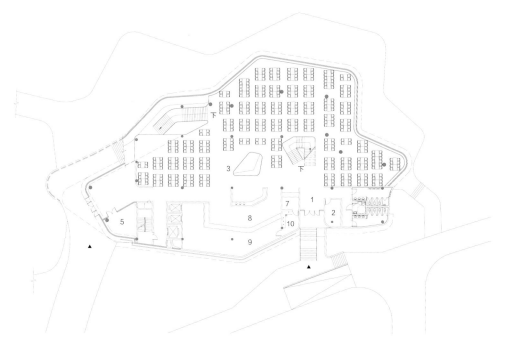

1 门厅
2 餐卡充值
3 就餐区
4 亲水平台
5 超市
6 备餐间
7 收残区
8 取餐通道
9 送餐通道
10 洗碗间

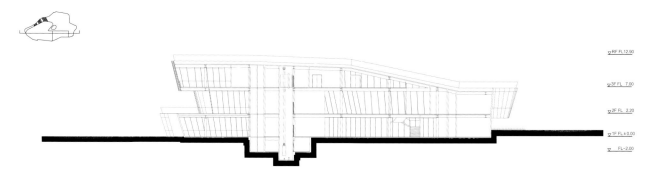

海盐杭州湾智能制造创新中心

海盐杭州湾智能制造创新中心

园区空间的多义性表达——海盐杭州湾智能制造创新中心创作体验

王振军 夏璐 徐彤

项目背景

海盐杭州湾智能制造创新中心位于国家中欧合作试点的浙江省海盐县，项目基地背靠海盐城面朝大海，而且离海岸最近处只有不到800m，因此它可能是目前中国距大海最近的园区，这也就成为该项目最大的区位特色。项目规划用地面积15.16公顷，总建筑面积26.95万m^2。

设计概念

基于项目面海靠城的区位特点和科技园区对高品质物理空间的目标追求，规划以"编织、交融"的概念，希望建筑与人、人与科技通过自然地交织，最终交融在滨海的自然脉络中。

在信息时代的背景下，信息共享变得前所未有的便捷，世界变得越来越开放，人们要求有更大的选择性和自主性。传统园区的规划模式已经不能为这样一个多元、复杂的社会需求作出回应。我们所面临的不仅是功能和技术上的挑战，更是观念上的改变。

现代园区设计中，公共空间常常被作为园区中必要的一部分，而将此空间与自然场力、景观设计相结合，则可以赋予园区更丰富的交流空间，营造更绿色的城市友好界面。我们现在展望的是鼓励人与自然交流、城市与海洋交流的复合型空间，希望通过"交融"的概念重新编织园区的规划结构，并使之与场地东南侧的海洋湿地公园相连接，使园区成为从城市向自然、向大海过渡的介质，打造具有生态、创新及前瞻性的新型智能园区。

城市秩序的重整、园区对创新氛围的激发、生态环境的延续、绿色建筑的打造，都是在开始设计该园区时首先要回答的问题。

设计目标

设计的目标和愿景既是宏大的也是具体的：

1. 人与自然通过科技的整合：希望城市与海洋、人、科技和自然在此交织，最终交融在自然的脉络中；

2. 一切从入驻企业的需求出发，使规划和建筑能灵活地适应各种招商模式的需要；

3. 快捷高效多功能集成的研发城：充分考虑入园企业生产、研发、生活需求，来配备、预留功能用地和建筑；

4. 最大化利用自然资源，以节省造价；

5. 发挥区位优势，打造海、城一体的开放园区，开发高科技旅游。

园区规划结构

1. 从城市肌理向海洋肌理的过渡：园区处于城市的秩序与大海的变化两种不同的肌理过渡中，如何使园区的肌理很自然地与城市和大海顺畅衔接、自然转换，是规划结构设计关键，也是契机。

2. 根据海盐风玫瑰图，顺应当地主导风向打造园区连续、通畅的生态廊道和通风廊道：通过底层架空、悬挑、门洞，打通园区空间和慢行系统，同时将通风廊道通过下沉庭院引入地下车库，以打造被动式的节能车库。

3. 园区与城市和大海的边界采用不同处理方法。

园区空间的建构及其多义性表达

毫无疑问，园区空间的表现力来源于其独特的秩序，同时也来源于其秩序框架下的多义性。秩序存在着不同的层级，比如从规划、单体、景观、材料、肌理、色彩、纹样等都是秩序建构框架下的具体层级和内容。

与一般位于城市或城郊结合带的园区不同，本项目具有位于大海和城市过渡带的区位特点，其自然场力和深层结构决定了园区的活力与秩序、自然与人工之间平衡的重要性，以及如何处理园区内部组团断裂与联系、围合与穿插，和单体的平衡与突变、简约与复杂、厚重与柔软等关系，以上这些要素之间的建构与平衡，无论在规划结构、空间塑造、单体设计、景观设计等方面还是在节能绿建设计方面，成为我们对该园区设计的出发点和归属。

建筑高度和体量由北向南、从城市向大海方向逐渐降低和缩小，使后排研发建筑顶层和屋顶均有良好的观海条件；公共服务空间及集中景观要素逐渐增加，建筑密度逐渐变小，呈现出一种从城市向自然的过渡状态。

园区重点建筑设计

1. 作为园区最基础单体类型的研发中试楼，其功能和空间灵活性是园区建筑最需具备的特点，由此引入服务空间与被服务空间的概念，为适应招商企业入驻创造最大的可能性。板式研发楼的布置和处理寻求在秩序中的突变，靠近城市一端追求与城市的衔接和呼应，在园区一端引入"风从海上来"概念，顶部结合屋顶平台做"漂浮"处理，仿佛徐徐的海风吹入园区，建筑与自然场力相辅相成，有机地交融在一起。

2. 作为园区中心和制高点的总部大楼，以V字形平面面向大海方向，试图争取最大化的观海面，并使北侧自然形成欢迎之态的主入口广场空间；而空间和形式的处理则通过左右两个体量的错动、穿插、咬合等手法追求建筑的动态平衡，在突出总部建筑大气恢宏气势的同时，表达我们在活力和创新方面的追求。

3. 位于"交织节点"上，柔性的餐厅点缀其中，其动线与园区流线融合在一起，采用非线性的形态是希望在节点位置更进一步激活园区空间的活力。餐厅的屋顶平台为员工们提供了更为丰富的公共交流场所。

空间节能为主、被动节能先行

在场地分析、总体布局、形态及空间生成、围护界面构造和材料选择等方面均充分考虑被动节能先行。一般园区地下车库的设计，大多采用的是全地下的方式。本项目综合考虑场地竖向和基础地质条件后，根据结构抗浮因素确定将建筑主体 ±0 标高确定在场地 1.45m，这样就有条件将庭院下沉到地下车库的标高。通过高侧窗和下沉式庭院的引入，使地下室白天完全可以靠自然采光解决照明问题。同时，下沉庭院与高侧窗为地下车库引入了新鲜空气，也被作为消防和排烟的通道。其他主要建筑采用大面宽、小进深的板式布局，充分体现被动式节能的设计追求。

结语

以上所述这些努力都是为了在大的秩序里通过不同层级的多义性表达来寻求空间体验的多样性，从而去激发园区空间的活力。借用一位同行对贝聿铭美国大气研究中心项目的评论，高科技园区就应该是"营造一种非机械性的、灵动的，并能启发新思想和贯通智能的无形气氛，让拥有不同学识的使用者能流连忘返或幽居独处，抑或是欣赏朝夕不同的美景之寄情所在。园区里代表人工作用结果的建筑应是来为扩展人类智慧、精神和美学追求而设计的，而不仅仅是一个普通的办公所在"。[1]

以上就是我们的思考、实践和追求，而要真正达到这种目标和境界，我们认为还有很长的路要走。

注解：

[1] 1985年2月28日沃尔特·奥尔·罗伯茨博士的口述历史记录提及全案的设计过程。

本文引改自《当代建筑》2023（04）

北京中关村移动智能创新园·Beijing Zhongguancun Mobile Intelligent Innovation Park

设计 Design 2014·竣工 Completion 2022

地点：北京·用地面积：6.99 公顷·建筑面积：34.30 万平方米
Location：Beijing·Site Area：6.99ha·Floor Area：343,000m²

合作单位：中国建筑设计研究院有限公司 一合建筑设计研究中心 U11 工作室
Cooperation unit：United Design 11 China Architecture Design & Research Group

合作设计师：陈珑、徐磊、张波、孙成伟、徐彤、鲍亦林、王舒越、邓涛
Co-designer：Chen Long，Xu Lei，Zhang Bo，Sun Chengwei，Xu Tong，Bao Yilin，Wang Shuyue，Deng Tao

该项目具有容积率高,且建筑高度受限的特点,因此呈现出高建筑密度的基本状态。规划采用合院布局,除南侧因住宅日照问题,高度略低,其余建筑均按规划限高设计。根据招商和使用效率需求,规划采用不同尺度的矩形单体,沿城市界面打造简洁、连续的建筑形象。而面向庭院一侧,强调界面的进退变化,结合不同尺度的庭院,打造富有活力的园区内部空间。由此形成建筑体量和规模的多样性,为不同互联网企业的入驻提供了多样选择。 规整的建筑形体为单体的设计带来挑战,为此我们采用具有不同质感的材料、色彩、细部、局部的开窗变化、裙房处理等手法,强调整个建筑群整体感下的差异化。同时,将原三元牛奶厂场地中的工业建筑遗存,如结构构架、工艺设备、符号作为设计语言保留了园区的场所记忆,在强调现代感的同时又关照了文化传承。灰色和红色的干挂陶砖,本色、酒红色、镜面铝板,辅以不同颜色的窗套及穿孔板,使建筑组群呈现出丰富的变化和活力。中心景观结合尺度庞大的地下服务空间和停车、人防系统做下沉处理,将充沛的阳光、新风引入地下空间。地上空间地下化,有效地减缓了地上建筑的压力。景观系统中,将原有工艺管线重新设计成不锈钢雕塑,画龙点睛地表明了项目的开发背景。

The project features a high plot ratio and restricted building heights, thus presenting a basic state of high building density. The planning adopts a courtyard layout, with slightly lower heights on the southern side due to residential sunlight issues, while the rest of the buildings are designed to the planned height limit. According to investment and efficiency requirements, the planning employs rectangular buildings of different scales to create a concise and continuous architectural image along the urban interface. Facing the courtyard side, emphasis is placed on the variation of interface advancement and retreat, combined with courtyards of different scales to create a vibrant internal space within the park. This results in diversity in building volume and scale, providing various choices for the entry of different internet enterprises. The regular architectural form poses challenges to the design of individual units. To address this, we employ techniques such as different textures, colors, details, patial window variations and podium treatments, emphasizing differentiation under the overall sense of the architectural complex. Simultaneously, the industrial architectural remnants in the original Sanyuan Milk Factory site, such as structural frames, process equipment, and symbols, are preserved as design languages, retaining the site memory of the park while emphasizing modernity and cultural heritage. Gray and red hanging ceramic tiles, original colors, burgundy, and mirrored aluminum plates, supplemented by window frames and perforated panels of different colors, create rich variations and vitality in the architectural complex. The central landscape integrates large-scale underground service spaces, parking, and civil defense systems, bringing abundant sunlight and fresh air into the underground space. The underground transformation of above-ground spaces effectively alleviates the pressure on above-ground buildings. In the landscape system, stainless steel sculptures redesigned from original process pipelines serve as the finishing touch, indicating the development background of the project.

北京中关村移动智能创新园

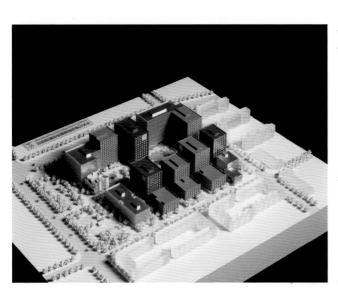

模型 / Model
总平面 / General Layout

潍坊崇文新农村综合服务基地·Weifang Chongwen New Rural Comprehensive Service Base

设计 Design 2015·竣工 Completion 2018

地点：山东，潍坊·用地面积：5.46 公顷·建筑面积：14.45 万平方米
Location：Weifang, Shandong·Site Area：5.46ha·Floor Area：144,500m²

合作设计师：孙成伟、陈珑、李达、邓涛、夏璐、鲍亦林、徐彤、张良钊、方雪、时菲、李洁
Co-designer：Sun Chengwei, Chen Long, Li Da, Deng Tao, Xia Lu, Bao Yilin, Xu Tong, Zhang Liangzhao, Fang Xue, Shi Fei, Li Jie

本项目为既有园区的扩建项目，规划结构采用"柔性组团"理念将旧有建筑与新建建筑组合在一起。每个组团强调建筑与景观、建筑与建筑之间的连续性，并围绕组团分中心强调其相对完整性，同时与组团分中心共同烘托园区中央景观，从而营造出具有整体感的空间结构。中央景观广场是基地最为重要的活动空间，它是园区慢行系统的枢纽，将来自园区各个方向的慢行步道组织联系在一起，同时作为园区视觉中心，园区各主要建筑的观景视线汇聚于此。位于园区中轴线上的、外凹内凸的总部大楼充分将自身融入中心景观区内，同时面向城市呈现出谦逊的姿态。建筑形象处理，面向城市强调整体理性严谨，面向中心景观强调变化通透，并通过退台架空将景观引入其中。将原有建筑立面重新设计，建筑群体在形态上呈现出一种理性的统一，同时又通过形体的变化和材料的选取强调单体的丰富性。建筑的色彩考虑整体风格典雅协调，外饰面以干挂陶板为主，返璞归真的质感与掩映在景观中的建筑相得益彰。

This project involves the expansion of an existing park, with the planning structure adopting the concept of "flexible clustering" to integrate existing and newly constructed buildings. Each cluster emphasizes the continuity between buildings and landscapes, as well as among buildings themselves. It revolves around the central cluster to emphasize its relative completeness while collectively enhancing the central landscape of the park, thereby creating a cohesive spatial structure. The central landscape plaza is the most important activity space on the site, serving as the hub of the park's pedestrian system, connecting slow trails from various directions within the park. Simultaneously, it serves as the visual center of the park, where sightlines from major buildings converge. The headquarters building, located on the central axis of the park, features a concave-convex design that integrates itself into the central landscape area while presenting a humble posture towards the city. The architectural expression emphasizes overall rationality and rigor towards the city while highlighting variability and transparency towards the central landscape, incorporating the landscape through terraces and elevated platforms. The redesign of the original building facades aims to achieve a rational unity in form, while also emphasizing individual richness through morphological variations and material selection. The consideration of architectural colors aims for an elegant and coordinated overall style, primarily utilizing dry-hung ceramic panels, complementing the natural texture with the buildings nestled within the landscape.

生长与蔓延 蔓·设计实践 I

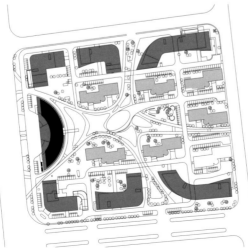

总平面 / General Layout

潍坊崇文新农村综合服务基地

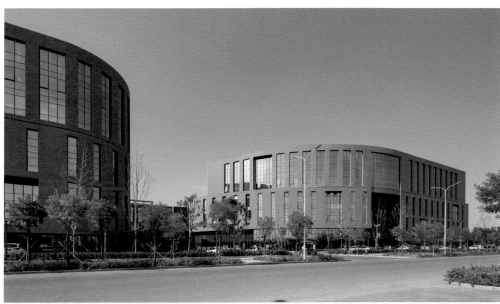

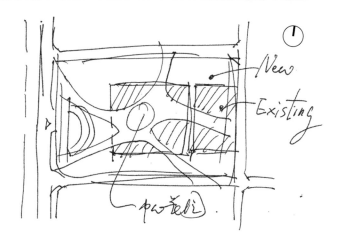

设计草图 / Sketch

北京大兴国际机场信息中心（ITC）、指挥中心（AOC）· The ITC and AOC of Beijing Daxing International Airport

设计 Design 2015 · 竣工 Completion 2019

地点：北京 · 用地面积：2.45 公顷 · 建筑面积：3.20 平方米
Location：Beijing · Site Area：2. 45ha · Floor Area：32,000m²

合作设计师：李达、陈珑、孙成伟、张良钊、赵玮璐、夏璐、王沛、邓涛、徐彤、申明、方雪、时菲
Co-designer：Li Da, Chen Long, Sun Chengwei, Zhang Liangzhao, Zhao Weilu, Xia Lu, Wang Pei, Deng Tao, Xu Tong, Shen Ming, Fang Xue, Shi Fei

本项目以北京大兴国际机场所处的显赫位置和本身具备的重要枢纽功能，成为新机场区域内仅次于航站楼的最重要和最具标志性建筑。设计用"信息链"的概念将信息中心（ITC）与指挥中心（AOC）连接起来，将技术、自然与人结合起来，向大家展示出该建筑智能、有机、人文的本质特性和大气、理性、质朴、可靠的建筑性格。建筑西侧尽可能退让紧邻的机场高架，将以指挥办公为主的AOC部分置于地块南侧，以获取最好的景观和朝向；以数据运维为主的ITC部分置于临近支路的北侧，二者相对独立并通过架空通廊和景观步道串联成一个整体；办公和研发体量以深灰色花岗岩塑造出连续环绕的体量节奏，强调安全的核心功能部用乳白色人造石板包裹嵌入，二者搭配散发出独特的人文气息；不同类型的景观庭院穿插其间，强调了高智能建筑崇尚自然以及对"以人为本"的设计原则的不懈诉求。ITC机房部分按照国际标准适度超前设计，一次施工、分部安装，便于后期扩展和机房等级提升，以充分考虑建筑全周期节能的可持续性。

Considering the outstanding position of this project in Beijing Daxing International Airport and its important pivotal function for the airport, it becomes a most significant and iconic building, second only to the terminal, in the new airport area. The design link the ITC and AOC via the concept of information chain, connecting technology, nature and people, so as to demonstrate the essential features of the building as being intelligent, organic, and people-oriented, and the architectural traits of being grand, rational, pristine and reliable. The west side of the building sets back as much as possible from the airport flyover nearby; the AOC, which is focused on command and office, is arranged on the south side of the plot so as to get the best view and orientation; the ITC, which is focused on data operation and maintenance, is placed on the north side, which is close to the access road. The two relatively detached parts are linked as a whole through elevated breezeway and landscape footpath; the dark grey granite applied on the office and R&D part generates a continuous surrounding mass rhythm; the core function part, which stresses security and safety, is embedded after being packaged with oyster white artificial stone, the collocation of the two materials gives off special cultural air; various types of landscape courtyards interleave among them, highlighting the advocating of nature in intelligent building as well as the unremitting appeal for the people-oriented design principle. Moderately forward-looking design is applied for the ITC mechanical room based on international standard. The construction of which is undertaken at one time while installation will be completed in steps, so as to facilitate subsequent expansion and upgrading of the mechanical room. Thus, full consideration is given to the sustainability of energy efficiency in the full period of construction.

北京大兴国际机场信息中心（ITC）、指挥中心（AOC）

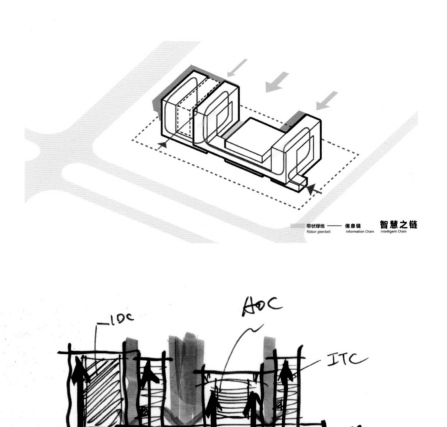

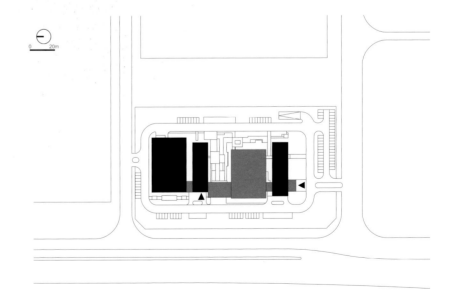

设计概念 / Design Concept
设计草图 / Sketch
总平面 / General Layout

生长与蔓延　蔓·设计实践 I

首层平面 / First Floor Plan
二层平面 / Second Floor Plan
剖面 / Section

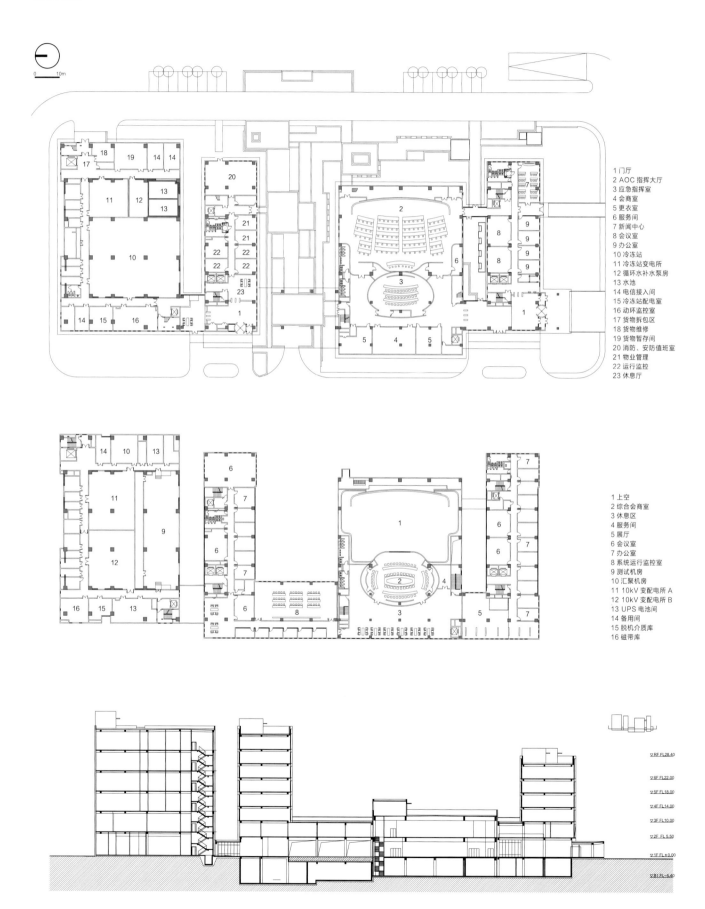

北京大兴国际机场信息中心（ITC）、指挥中心（AOC）

概念生成 / Concept Generation
模型 / Models

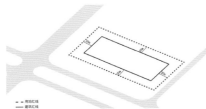

Step 1 项目用地：建筑项目规划用地 24500m²，由于红线内有管线，建筑退让用地红线 20m。

Project land use: 24500m², the construction sets back 20m from the property line as there are pipelines underneath.

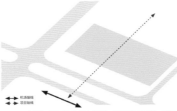

Step 2 轴线：根据机场总体规划，确定机场轴线。根据机场轴线，确定项目轴线。

Axis: The airport axis is determined by the master plan of the airport. The project axis is determined as per the airport axis.

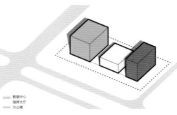

Step 3 功能布局：根据任务书要求，布置数据中心、指挥大厅和办公楼体块。

Functional layout: The ITC, AOC and office building masses are arranged as per the requirement of the project program.

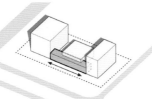

Step 4 连接：通过休息及共享空间连接数据中心、指挥大厅与办公空间。

Connection: The ITC, AOC and office space are connected through rest and shared spaces.

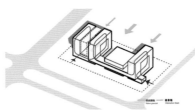

Step 5 信息链：概念植入建筑形体中，并插入中庭。

Information Chain: The concept is embedded into the architectural form, with an atrium incorporated into the building.

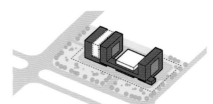

Step 6 完成：置入道路与绿化，完成项目。

Completion: embedding roads and greening to conclude the project.

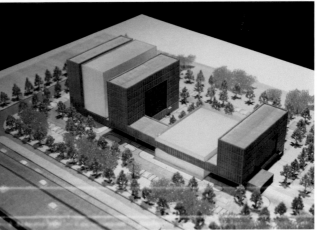

生长与蔓延　蔓·设计实践 I

190

北京大兴国际机场信息中心（ITC）、指挥中心（AOC）

北京大兴国际机场生活服务设施工程 · Beijing Daxing International Airport Life Service Facilities Project

设计 Design 2016 · 竣工 Completion 2019

地点：北京 · 用地面积：3.23 公顷 · 建筑面积：6.60 万平方米
Location: Beijing · Site Area: 3.23ha · Floor Area: 66,000m²

合作设计师：孙成伟、徐彤、陈珑、李达、郑秉东、王沛、邓涛、赵玮璐、王舒越、时菲
Co-designer: Sun Chengwei, Xu Tong, Chen Long, Li Da, Zheng Bingdong, Wang Pei, Deng Tao, Zhao Weilu, Wang Shuyue, Shi Fei

本项目提出"绿色·交融"的设计概念,希望为使用者提供一个放松、便捷的复合型空间环境。设计中尝试将倒班宿舍、生活服务中心、职工食堂等功能重新组织,以围合的方式进行垂直分区。其中下部生活服务中心及职工食堂部分引入开放街区的概念,为机场6000人提供便捷的就餐和购物空间的同时,利用屋顶花园为倒班人员创造更多的休闲、运动空间,并最大限度地将场地南侧优美的自然景观资源引入到场地之中,提升室内与庭院的空间品质;上部的轮班宿舍则采用合院式布局围合出完整的体量,其中南北侧为单廊南向宿舍布局,东西侧为单廊双向布局,以保证每间宿舍均有阳光;同时,将靠近机场高架一侧的房间处理为封闭阳台,以积极回应场地高速路噪声的影响。立面材料选用浅灰色涂料搭配金属深窗套,简洁利落、美观节能并易于维护。整个设计期望打造人与自然交融、建筑与自然交融、城市与建筑交融的,具有生态、创新、便捷的生活服务设施综合体。

With the "Green·Blending" design concept proposed for the project, it aims to provide the users with a relaxing and convenient composite space environment. In the design, it tries to reorganize the functions of shift dormitories, living service and staff canteen, making vertical partitioning by means of enclosure. The lower section, which includes the living service center and staff canteen, introduces the concept of an open street, offering a convenient dining and shopping space for 6,000 airport employees. Additionally, a rooftop garden provides extra leisure space for shift staff. The natural landscape on the south side of the site is maximally integrated into the plot, enhancing the quality of the interior and courtyard spaces. The upper-level shift dormitories are organized in a courtyard layout, forming a cohesive structure. The north and south sides feature single-corridor, south-facing dormitories, while the east and west sides employ a single-corridor, dual-aspect layout to ensure that each dormitory receives adequate sunlight. Rooms located closest to the airport flyover are fitted with enclosed balconies to mitigate noise from the expressway. For the facade materials, light gray painting plus metal thick window frames are used as they are simple, beautiful, energy-efficient, and easy to maintain. It aims to create an ecological, innovative, and convenient living service facilities complex that blends people with nature, buildings with nature, and city with buildings.

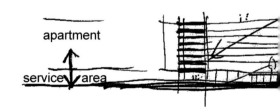

总平面 / General Layout
设计草图 / Sketch

生长与蔓延　蔓·设计实践 I

首层平面 / First Floor Plan
标准层平面 / Typical Floor Plan

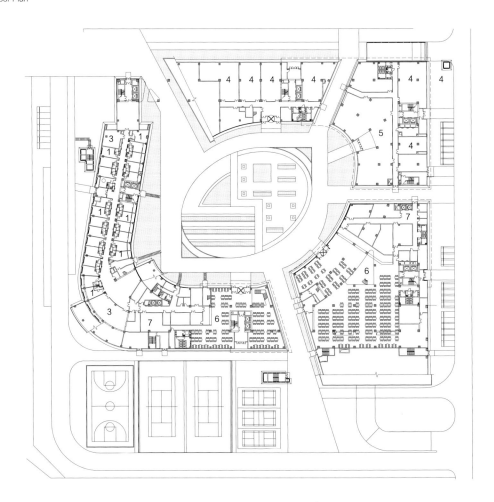

1 员工宿舍
2 管理人员宿舍
3 活动室
4 生活服务中心
5 咖啡厅
6 餐厅
7 厨房
8 休闲区
9 室外平台

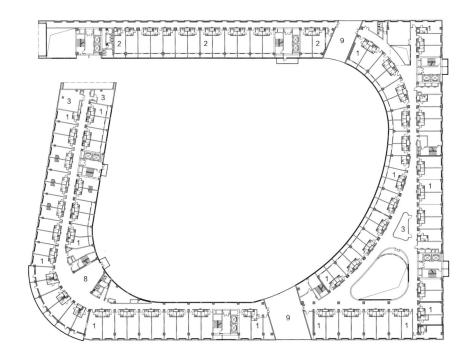

北京大兴国际机场生活服务设施工程

模型 / Model

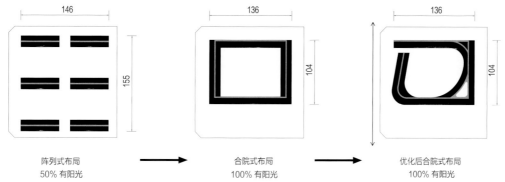

阵列式布局　　　　　　合院式布局　　　　　优化后合院式布局
50% 有阳光　　　　　　100% 有阳光　　　　　100% 有阳光

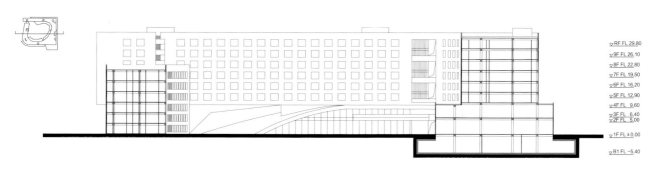

布局分析 / Layout Analysis
剖面 / Section

北京大兴国际机场生活服务设施工程

阳光之城——北京大兴国际机场生活服务设施工程设计

王振军　徐　彤　何　霈

摘要：北京大兴国际机场生活服务中心项目，通过全部公寓均为南向、东向和西向的布置方式，以及公寓建筑形体的架高、压低等手法，实现了"每个员工都能沐浴阳光"的设计目标；并利用合院式的建筑布局，及多处嵌入式的活动空间，创造出了生动的空间体验和丰富的运动场所，最终实现提升空港员工的生活品质的愿景。

即使在科学技术高度发达的今天，我们的生命健康仍然不断遭受挑战。保证人类个体健康、平安的要素有很多，而"沐浴阳光"无疑是当中最基本、最重要的要求。本项目是北京大兴国际机场配套的、可容纳1700余名员工的集体公寓。方案构思是从营造一个绿意、阳光的宿舍设想开始的。地处寒冷地区的北京，整体建筑形态呈现扁平化水平延展的大兴国际机场，"让这里的每位员工都能沐浴阳光"是设计展开的第一要义，然后才是其他方面的考量，如生动的空间体验、丰富的运动场所、完善的配套设施等。

1 空间制约下的总体布局

本项目地处北京大兴国际机场服务区内，特殊的用地区位，决定了整个场区的建筑都具备着庞大的建筑体量和严格受限的建筑高度。为应对特殊场区内的建筑特征，同时为了实现公寓间间有阳光的目标，本项目没有简单地应用传统公寓建筑的阵列式布局，而是选择了具有连续界面的合院式布局，达到了整合建筑体量、重塑室外活动场所的目的。在建筑形体上，将地上面积为52100m²的建筑体量以连续的合院式布局呈现，呼应了机场场区内建筑的大体量特征。在空间营造上，围合的建筑形式，形成了完整的内庭院休闲场所，相较于分散的阵列式布局，围合的建筑形式在建筑占地上更加集约，为场地南侧运动场地的腾退创造了条件。在场地南侧，日照充分的位置，共布置了标准篮球场1个，标准网球场2个，标准羽毛球场3个。合院内外的休闲运动场所，为居住在这里的年轻人提供了高品质的活动空间，为他们日复一日的严谨、紧张的工作增添了一抹色彩。

2 内化与外向的空间营造

围合的建筑在西北角向城市打开，形成整个公寓的主入口广场。架空及悬挑的上部建筑，以及从地面倾斜上升的裙房绿色种植屋面，为入口广场提供了丰富的空间体验和荫凉、绿意的使用感受。同时，建筑裙房的东、南、北三向均设有内外贯通的开口空间，增强了内院的通达性，为公寓提供了丰富的中介空间。

从空港的城市空间到园区空间，从入口的中介空间到围合的建筑内庭空间，再到由不同尺度和开放度组成的内部空间，富有层次的空间营造为员工带来了丰富的空间体验。

项目场地西侧紧邻机场出场高架路，建筑形体在此处做压低处理，既丰富了合院的天际线，突出了主要展示角度的建筑形象，完善了"看与被看"的视觉感受，又将充沛的阳光引入建筑内庭院，更好地激活了庭院的休闲空间。

在尺度上，连续界面中用灰空间来展示建筑的空隙感和深度；在体量上，面向街区的体量并没有太紧凑，地面更加通透，以实现视线水平面上更好的视觉联系。我们还尝试以丰富的界面激活机场周围规整化的布局。公寓部分立面采用白色涂料，架在半透明的裙房之上，深窗框施以黄色，象征着活力与阳光。

3 功能为先的竖向分区

为提升服务的便利性和北侧公寓的日照品质，该项目两层高的裙房部分被设计成以餐饮、零售商业、健身等功能共享的公共空间。裙房的玻璃幕墙为这些公共空间提供了更好的采光，也将中心庭院的景观最大限度地引入室内。这样所形成的透明界面，以及地面和屋面有机的过渡的处理，创建出了一种建筑内、外部环境之间的友好关系。

建筑三层以上均为公寓。为使每一间公寓都可达到"沐浴阳光"的目标，在平面布局的选择上完全杜绝了不见阳光的纯北向房间，所有房间的外窗均朝南、东及西三个方向，真正实现了户户有阳光。同时，考虑到东侧、西侧的太阳高度角小，为避免过多的太阳辐射对室内舒适度产生影响，设置了封闭阳台或突出的窗套，并在室内加装高反射率窗帘，使得所有公寓房间都尽可能实现既能照射到阳光，又可根据需求调节光照强度的目标。

与此同时，本项目还提出了一种我们认为适宜当代空港区的居住形式，着力营造充分的共享交流空间，以及镶嵌于建筑中的不同尺度的采光中庭和灰空间，借助多样化的公共空间来弥补机场由于宏大体量所造成的场所缺失。阳光可以通过多种方式进入建筑内——或通过落地的外窗，或通过代表活力的黄色深窗框，抑或通过建筑内不同尺度和方向的采光天井，多种途径共同打造随时间而变换的光影效果。公寓单元也为业主提供了单人间、双人间、三人间和四人间，以及顶层跟随层高变化而设计的复式单元，以满足居住的各种标志性需求。

4 绿色建筑实践

户户有阳光的设计初衷和重点打造的丰富公共活动、服务空间奠定了整个公寓建筑的高品质基础，同时该项目继续挖掘绿色建筑策略，以求最大限度地实现建筑的可持续发展。目前，生活服务中心已获得中国绿色建筑三星认证。

本项目在裙房部分设置了倾斜的屋面及屋顶绿化，屋顶绿化面积达到 4022.4m^2，占屋顶总面积的 35.93%。屋顶绿化可有效地增加场地绿化层次，增加空气湿度，吸附尘埃，减少噪声，并降低室内的热辐射，节约空调能耗，为使用者创造更加舒适、宜人的室内外环境。同时，屋顶绿化可吸收并利用天然降水，减少地面排水压力。本项目通过挡土板＋防滑格的构造方式，成功解决了倾斜屋面的种植问题，这是我们对于绿化屋面的新尝试。利用起伏的屋面，为员工设计了 400m 长的屋顶慢跑跑道，以此形成富有趣味的慢行系统运动空间。郁郁葱葱的植物，围合出内部庭院为员工提供娱乐、社交及亲近自然的场所。

公寓的建筑性质决定了生活热水的需求极大。因此，本项目充分利用太阳能资源，在屋顶设置 386 套，共 1554m^2 的太阳能集热板，供应公寓卫生间的淋雨及洗手盆热水。生活热水总用量为 130.4t，可再生能源供应比例可达到 76.33%。

除此之外，北京大兴国际机场具备全球最大的浅层地源热泵集中供能系统。地源热泵系统地埋管集中布置于机场蓄滞洪区内，为机场功能用房及驻场单位用房提供制冷和供热服务，总供热能力 176kW，总供冷能力 123kW。依据北京气候及项目用能计算（冬季供暖期 123 天，夏季制冷期 120 天，日使用时间 24h），可满足 2570000m^2 建筑的冬季供暖和夏季制冷需求。本项目的冷热源充分利用机场的地源热泵系统，输配系统选用变频水泵，采用风机盘管＋新风的空调末端，风机盘管暗装在吊顶内。

本项目在食堂、咖啡厅和商店采用新风机来提高室内空气品质，在新风机内部设置新风热回收。在冬夏两季，室内、外温差较大时，通过排风能对新风进行有效的预冷、预热，节约能耗。本项目利用非传统水源，进行室内冲厕、室外绿化灌溉、道路浇洒、车库冲洗等，建筑可回用水量达 35882.74m^3/天。

5 结语

通过该项目的设计实践，我们更加深刻地感受到：建筑的意义，是通过建造活动连接人与自然、连接人与人、连接社区肌理，从而参与构筑最基本的文明条理的。

正如罗马大学 AnnaIrene Del Monaco 的"Architecture and Communities Today"文中所说："当今社会期待建筑扮演的角色之一，是能够塑造社区，重塑环境，并以一种可预测的、积极的方式影响人的行为。此外，建筑师自认为拥有的职业技能，就是通过建筑概念和设计来呈现社会的想象和希冀。"[1]

我们在北京大兴国际机场生活服务中心的这一设计实践中，试图通过建筑设计去激发空港社区的活力，最终提升空港员工的生活品质。

参考文献：

[1] 莫纳科，黄华青. 今日的建筑与社区 [J]. 世界建筑，2020，356（2）：24-29.

北京大兴国际机场工作区车辆维修中心工程·Beijing Daxing International Airport Work Area Vehicle Maintenance Center Project

设计 Design 2016 · 竣工 Completion 2019

地点：北京 · 用地面积：1.55 公顷 · 建筑面积：6568 平方米
Location: Daxing, Beijing · Site Area: 1.55ha · Floor Area: 6,568m²

合作设计师：陈珑、郑秉东、李达、孙成伟、赵玮璐、徐彤、邓涛、时菲、李洁
Co-designer: Chen Long, Zheng Bingdong, Li Da, Sun Chengwei, Zhao Weilu, Xu Tong, Deng Tao, Shi Fei, Li Jie

北京大兴国际机场（简称大兴机场）工作区车辆维修中心是机场大量特殊车辆的维修基地，为机场运行提供有效保障。根据车辆尺度，将维修车间相应位置屋面朝南向掀起，形成采光通风天窗，为维修厂房提供良好的光照条件，基于功能的倾斜的屋面使建筑展现出机场建筑特有的韵律感和动感，并有效地消解了建筑扁平的尺度。结合庭院，维修车间与办公辅助楼组合成 U 形组群，既相互独立又联系便捷。灰白色彩组合更加强调建筑在机场辅助区作为背景建筑的功能特性。

Beijing Daxing International Airport Work Area Vehicle Maintenance Center Project serves as a repair base for a large number of specialized vehicles, providing effective support for airport operations. According to the scale of the vehicles, the roofs of the maintenance workshops are lifted to the south to create skylights for lighting and ventilation, ensuring favorable lighting conditions for the maintenance facilities. The functionally inclined roofs give the building a distinctive rhythm and dynamism characteristic of airport architecture, effectively mitigating the flat scale of the building. Combined with courtyards, the maintenance workshops form a U-shaped cluster with office auxiliary buildings, which are both independent and conveniently connected. The gray-white color combination further emphasizes the functional characteristics of the buildings as background structures in the airport auxiliary area.

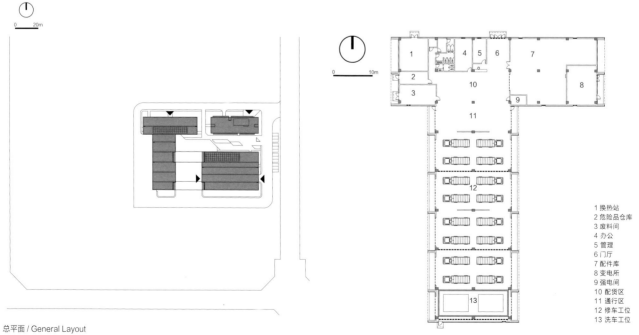

1 换热站
2 危险品仓库
3 废料间
4 办公
5 管理
6 门厅
7 配件库
8 变电所
9 强电间
10 配货区
11 通行区
12 修车工位
13 洗车工位

总平面 / General Layout
首层平面 / First Floor Plan

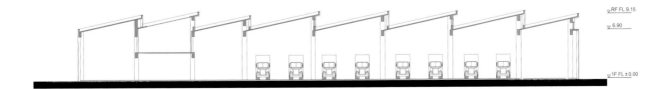

剖面 / Section

生长与蔓延　蔓·设计实践 I

北京大兴国际机场工作区车辆维修中心工程

中国残联北京按摩医院扩建项目 · China Disabled Person's Federation Beijing Massage Hospital Expansion Project

设计 Design 2017 · 竣工 Completion 2022

地点：北京 · 用地面积：1.03公顷 · 建筑面积：3.99万平方米
Location：Beijing · Site Area：1.03ha · Floor Area：39,900m²

合作设计师：陈珑、李达、孙成伟、邓涛、鲍亦林、王舒越、郑秉东、朱谞、徐肜、邓涛、李洪鹏、车爱明、时菲、王凯、刘澈
Co-designer：Chen Long, Li Da, Sun Chengwei, Deng Tao, Bao Yilin, Wang Shuyue, Zheng Bingdong, Zhu Xu, Xu Tong, Deng Tao, Li Hongpeng, Che Aiming, Shi Fei, Wang Kai, Liu Che

北京按摩医院新址位于北京市朝阳区武圣北路。新院的建设全面致力于满足和适应医院未来发展需求，成为我国面向世界弘扬中医文化和展示对弱势群体关怀的人权事业的重要窗口。面对这一人文历史背景悠久且具有特殊意义的医疗建筑，需要打破传统医院设计模式，突出以人为本的核心价值，采用被动式绿建手法，遵循阳光医疗、景观医疗、绿色医疗的设计原则。在设计过程中提炼中国传统建筑的"檐""梁""柱"经典元素，将其转译并抽象成新的建筑语汇，与建筑功能有机地融合在一起，建筑与景观细部点缀应用了斗栱、月亮门、屏风、漏窗等传统要素；同时，建筑色彩以北京传统建筑中常见的深灰、浅灰色与原木色为主要建筑用色，体现了整个城市肌理在本项目中的传承与延续，最大限度地唤起按摩医院特有的地域与场所精神。设计将抽象后的中国传统建筑要素重塑与整合，使形式追随功能，实现有机建构，从而打造真正体现人文关怀的，既有传统意蕴，又具中国特色的医疗建筑。所谓当代，就是要重新利用过去，要将过去植入现在。

The new site of the Beijing Massage Hospital is located on Wusheng North Road, Chaoyang District, Beijing. The comprehensive construction of the new hospital is dedicated to meeting and adapting to the future development needs of the hospital, becoming an important showcase window for China to promote traditional Chinese medicine culture and show our concern for vulnerable groups in the field of human rights globally. Given the rich cultural and historical significance of this medical building, it is essential to break away from traditional hospital design models and emphasize the core value of a human-centered approach. A passive green building strategy should be adopted, following the principles of sunlight-based therapy, landscape-integrated therapy, and green therapy. The design refines the classic elements of "eaves" "beams" and "columns" of traditional Chinese architecture, translates and abstracts them into a new architectural vocabulary, and organically integrates them with the building functions. Traditional elements such as brackets, moon gates, screens, and louver windows are applied as embellishments in architecture and landscape details. At the same time, the architectural colors use the common deep gray, light gray, and natural wood colors found in traditional Beijing architecture as the main building colors, reflecting the inheritance and continuation of the entire urban texture, and maximally evoking the regional and contextual spirit unique to the massage hospital. This approach aims to create a medical building that truly embodies human care, evokes traditional aesthetics, and reflects distinct Chinese characteristics. The so-called contemporary is to reuse the past, to implant the past into the present.

生长与蔓延　蔓·设计实践 I

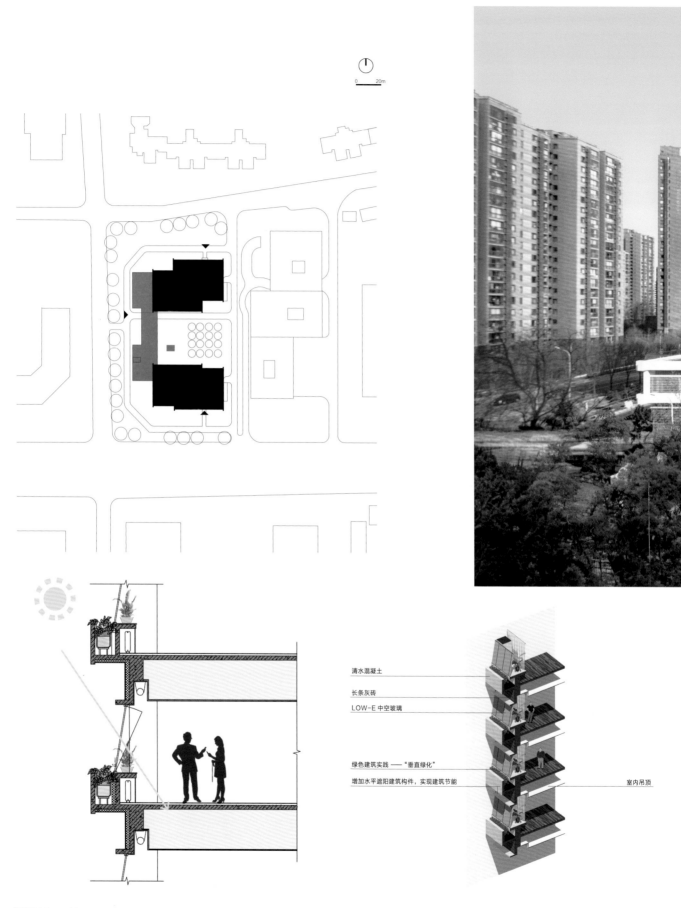

清水混凝土
长条灰砖
LOW-E 中空玻璃

绿色建筑实践——"垂直绿化"
增加水平遮阳建筑构件，实现建筑节能

室内吊顶

总平面 / General Layout
绿色技术 / Green Technology

210

中国残联北京按摩医院扩建项目

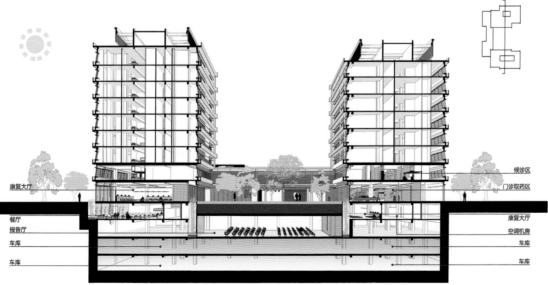

经典的U形布局围绕着疗愈花园，使门诊和住院空间都能充分地享受到阳光和景观。由于场地紧张，绿化率要求又很高，故在利用地下边庭引入阳光和自然风的前提下，充分利用地下空间布置医技功能、医疗街、康复中心以及员工餐厅、学术报告厅等功能。

剖面 / Section

生长与蔓延　蔓·设计实践 I

中国残联北京按摩医院扩建项目

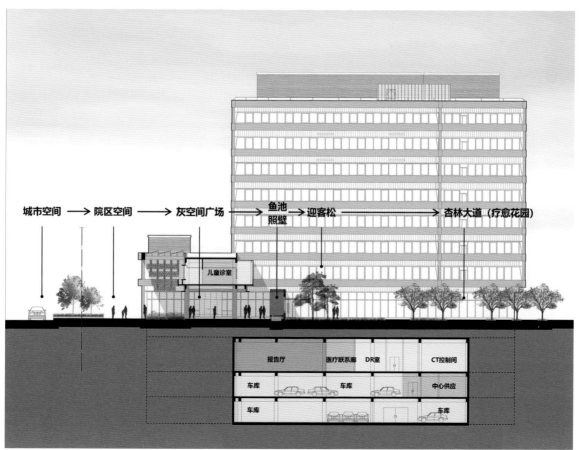

空间序列 /Spatial Sequence

生长与蔓延 蔓·设计实践 I

首层平面 / First Floor Plan
二层平面 / Second Floor Plan

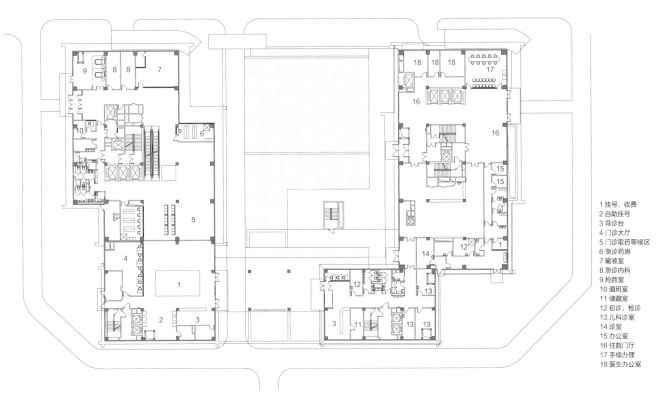

1 挂号、收费
2 自助挂号
3 导诊台
4 门诊大厅
5 门诊取药等候区
6 急诊药房
7 输液室
8 急诊内科
9 抢救室
10 值班室
11 储藏室
12 初诊、检诊
13 儿科诊室
14 诊室
15 办公室
16 住院门厅
17 手续办理
18 医生办公室

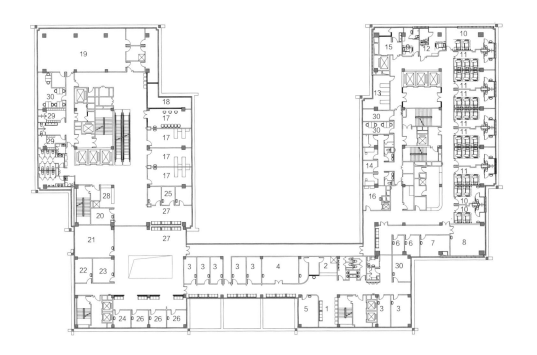

1 挂号、收费
2 号巡台
3 儿科诊室
4 针灸室
5 针刀室
6 ST 理疗室
7 OT 理疗室
8 PT 理疗室
9 单人病房
10 双人病房
11 三人病房
12 无障碍病房
13 按摩大厅
14 值班室
15 污物间
16 配膳间
17 治未病诊室
18 采血间
19 中心检验室
20 耳鼻喉科诊室
21 口腔科诊室
22 皮肤科诊室
23 眼科诊室
24 骨伤科诊室
25 内科诊室
26 外科诊室
27 候诊大厅
28 护士站
29 更衣室
30 办公室

中国残联北京按摩医院扩建项目

标准层平面 / Typical Floor Plan
模型 / Model

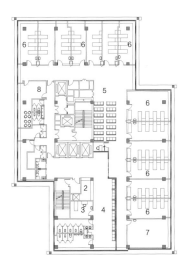
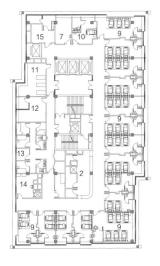

1 收费
2 护士站
3 初诊
4 一次候诊
5 二次候诊
6 普通按摩诊室
7 中医治疗室
8 示教室
9 病房
10 无障碍病房
11 按摩大厅
12 办公室
13 值班室
14 配膳间
15 污物间

生长与蔓延　蔓·设计实践 I

中国残联北京按摩医院扩建项目

形式与精神——中国残联北京按摩医院创作体验

王振军　陈珑　鲍亦林　王舒越

所谓当代，就是要重新利用过去，要将过去植入现在。

——题记

一　项目背景与定位

北京按摩医院原址始建于1958年，位于北京西城区宝产胡同，系在传统合院基础上改造而成，传承中国推拿技艺的同时，营造了轻松、舒适、宜人的就诊环境。自建成以来，多次承接了外交部安排的外交接待活动，为展示我国残疾人事业作出突出贡献。北京按摩医院新址位于北京市朝阳区武圣北路，建设用地面积为10289.0m^2，总建筑面积39899.0m^2。东临朝阳区综合医疗服务中心，南至南磨房路，西至武圣北路，北至八棵杨南街。西侧邻近地铁10号线，交通十分便利。

新址虽然造价有限，但对它的定位较高且有如下明确的要求：

1. 中国第一个按三甲标准设计的中医按摩康复医院；
2. 向世界展示中国残疾人事业的窗口单位（文化交流和人权事业展示）；
3. 中国残联无障碍标准示范工程。

二　建筑的复合性及其二元选择

根据项目的定位，我们不难勾勒出北京按摩医院无论是在形式、内容，还是在建筑精神表达上应该有如下侧重：

1　中国特色与时代精神

充分研究和发掘当地文化和精神的物质载体，通过建筑语言的独创性表达，塑造建筑的鲜明个性。在探索中国特色的同时，充分利用当代先进的建筑设计理念和成熟的建筑技术，呈现建筑的时代特色。

2　中国中医文化和中医文化建筑

注重中医文化建筑的塑造，营建中医文化建筑亲切、宜人的空间氛围的同时，推动我国医疗事业和人权保障事业的发展，使中国残联北京医院成为向世界进行文化交流与人权展示的窗口和空间场所。

3　以人为本

突出"以人为本"的核心理念，更加注重医患空间的舒适性、便捷性、无障碍通达性，以及按摩医院就诊患者的行为特殊性。

4　绿色建筑

设计中采用节能环保等新技术、新工艺、新材料，充分利用自然通风和采光，塑造生态绿色建筑。从自然环境中汲取设计要素，最大化地利用自然资源。既要达到节能的目的，又要节省一次性投资。

现代与传统、形式与内容、仪式感与便捷性、节能的主动与被动等的二元选择都是该项目应该回答的问题。

三　设计愿景

1　生态化环境

生态环境与人体健康息息相关。因此，减少对环境的污染、节约能源，进行人性化的设计对按摩医院来说，是至关重要的。

2　艺术化建筑

形式与功能完美结合，艺术化的建筑造型体现了医院与中国传统建筑的性格，在减轻就诊患者心理负担的同时，增强其对传统文化的认同感，并注重向世界展示中国传统建筑的精神。

3　人性化空间

针对按摩医院特殊需求，同时为使用者与患者提供更加便捷的流线，以及更加轻松、舒适、安全的公共空间；对于公共空间，特别是地下公共空间，通过自然通风、日照与景观的引入，提升室内环境品质。

四 设计理念

面对北京按摩医院悠久的人文历史背景及特殊的功能需求，传统医院建筑设计模式已经不能为这么一个多元化、特殊化的项目提出答案。

我们尝试提炼中国传统建筑的"檐""梁""柱"经典元素，转译并抽象成新的建筑语汇，同时与建筑功能有机融合，最终设计了具有创新性、前瞻性及可持续性的现代中医医院建筑。我们希望通过"传承与提升"的设计概念最大限度唤起按摩医院特有的地域与场所精神，在继承中国传统文化的基础上，又能体现传统中医在当代的发展与提升。

建筑语言：受艾森曼（P.Eiseman）将艾弗拉姆·诺姆·乔姆斯基（(Avram Noan Chomsky）等的符号学应用到建筑学上的启发（乔姆斯基的转换生成语法理论），通过对中国传统建筑的研究来探索在当代建筑建构时中国精神的表达。中国传统建筑元素的提取和转译，并抽象成现代、简约与纯粹的建筑语汇形式。

1 檐

屋檐是中国古典建筑立面最重要的构成要素，屋檐出挑的深远使建筑物产生独特而强烈的视觉效果和艺术感染力。重檐屋顶通过形式组合，调节建筑比例与层次，使建筑形体和轮廓变得愈加丰富，又使中国建筑的"第五立面"最具魅力。

2 梁

梁是中国传统建筑构架中最重要的构件之一，承托着建筑物上部构架及屋面的全部重量，是建筑构件中最重要的部分。传统的梁架体系有抬梁式和穿斗式两种主要形式，其中北方广泛采用的是抬梁式结构体系，特点是在柱头上插接梁头，梁头上安装檩条，梁上再插接矮柱用以支起较短的梁。

3 柱

中国传统建筑的柱是一种直立而承受上部荷载的构件，是中国传统建筑中最重要的构件之一，其中"角柱"位于建筑物的转角处，承托不同角度的梁坊。角柱的设置有利于增强结构的稳定性，勾勒建筑的立面轮廓。

将抽象后的中国传统建筑要素重塑与整合，通过整体塑造、局部建构（如入口灰空间）以及细部处理，最终实现整个建筑组群的有机建构。

用中国传统建筑的语言重构形成的整个建筑所特有的檐廊空间意向，形成了建筑与城市、建筑与环境自然渗透的空间层次，并有效改善了建筑对直射阳光的遮挡。

五 总图布局

南北向近东西向两倍的矩形用地决定了医院住院部和门诊楼的标准层平行布置于南北边界，并且尽量拉开距离，以获得最佳日照，而两者之间形成入口灰空间广场。

采用经典U形布局，也是考虑到未来东侧扩建的可能性。扩建完成后可形成更为安静的合院布局。

六 功能组织

经典的U形布局围绕着疗愈花园，使门诊和住院空间都能充分地享受到阳光和景观。由于场地紧张，绿化率要求又很高，故在利用地下边庭引入阳光和自然通风的前提下，充分利用地下空间布置医技功能、医疗街、康复中心以及员工餐厅、学术报告厅等。标准层布置优先满足住院病人和就诊患者对阳光的需求，病房均布置在南侧将北侧用房用于医护及辅助功能。

七 交通系统

本项目周边自身形成环路交通，结合医院内的机动车路道和市政路道形成环形消防通道，与医院的各主要交通道路连通，满足消防扑救要求。机动车的主要入口设在西侧。南北侧均设有机动车辅助出入口。

本着"人车分流、高效便捷"的原则，车行道路围绕场地周边布置。本项目主要出入口位于建筑的西面和南面，北面为住院部出入口。西出入口为门诊的主要出入口，南出入口为急诊出入口，北出入口为住院部探视出入口，北面靠东侧设置一个污物出口。

建筑首层设置了发热门诊、儿科、办公等出入口。主要门诊人流设置在西面，急诊人流设置在南面，探视人流设置在北面，办公人流设置在建筑的东面。在U形平面上有效组织了门诊、医技、住院的主要流线。

八　空间秩序

把原有北京按摩医院的灰砖移建至新址月亮门处，从而把原有北京按摩医院场所精神"移植"到新的建设场地中。实现新老建筑跨越时空的对话，把按摩医院的历史及其未来有机地联系起来。建筑既尊重了历史，也赋予了医院在区域内的新身份。

九　形象设计

在建筑立面的处理上主要考虑医院建筑需要宁静、轻松的氛围，以白色调为主，现代风格的水平长窗能最大限度的引进阳光及庭院景观，配合水平的遮阳百叶，使得各建筑体量得到统一，并在建筑立面上形成阴影，呼应北京的城市灰色主色调。木百叶及灰瓦坡屋顶可隐藏放置在屋顶的空调室外机，同时展现北京传统建筑的缩影，使之成为京城地区的精神归属与城市象征。

十　绿色建筑：被动优先、空间节能优先

地下边庭的引入激活了地下空间——严苛规划条件下地下空间的"地上化"。室外挑檐下植入的边庭将上下两层空间贯通，其正面的垂直挡墙使光线通过多重折射蔓延进地下功能空间，实现良好的自然采光和通风。与城市空间有机融合的灰空间广场既体现了礼仪性又表达了人文情怀。在表现建筑传统精神的同时又具备建筑遮阳功能。

十一　材料运用

本项目运用清水混凝土、木板（帕莱斯板）、木饰面百叶、灰色陶土面砖为主要材料并进行组合。

与紫禁城金碧辉煌的帝都气派不同，北京胡同、老街和四合院中所体现的灰瓦、灰墙与原木色构件奠定了北京城市色彩的主基调，灰色与原木色返璞归真，是北京最常见、也是最有中国建筑神韵的色彩。透出古都浑厚、朴实、宁静的文化底蕴。室内大量木饰面的使用，拟在打造宁静、疗愈的环境。

十二　无障碍设计

本项目的无障碍设计有其特殊性，其一是服务人群的特殊性：多为行动不便的老年人，需考虑轮椅的回转、轮椅的可达性等无障碍措施。其二是从业人群的特殊性：约有一半医护人员为视障或视弱人士，需考虑盲道、语音播报等无障碍措施。

本项目采用保障为主、通用为主、适用为主、体验为主的无障碍设计理念，以及逢棱必圆、逢台必坡、逢高必低、逢陡必缓、逢滑必涩、逢沟必平、逢隙衔接、逢源左右等设计原则，努力成为全国残联向世界展示我国人文关怀、向国内推广无障碍设计理念和方法的示范窗口，并成为执行国家标准《无障碍设计规范》GB 50763—2012 的示范工程。

本项目已入选第16届中国信息无障碍论坛——暨全国无障碍环境设计成果：无障碍环境建设优秀典型案例设计类优秀案例。

结语

传统精神的当代表达、中华文化及中医文化的彰显、医院人文精神与现代医学科学有机结合、绿色建筑中的被动与

主动等课题都是该项目研究和实践的重点。中国残联北京按摩医院建筑所涉及的典型的"复合性"是当代中国建筑师经常面临的课题，也是"中国式现代化"在建筑学领域的一个值得永远探究下去的永恒课题，需要我们在未来的实践中不断努力，去寻找答案。

本文引改自 AT 公众号 2023 年 8 月

北京大兴国际机场非主基地航空公司生活服务设施工程·Beijing Daxing International Airport Non-Main Base Airline Living Service Facilities Project

设计 Design 2017 · 竣工 Completion 2019

地点：北京 · 用地面积：3.12 公顷 · 建筑面积：8.42 万平方米
Location：Beijing · Site Area：3.12ha · Floor Area：84,200m²

合作设计师：陈珑、徐彤、孙成伟、申明、郑秉东、夏璐、王舒越、赵玮璐、时菲、李洪鹏、王凯、刘澈、王沛、周宝昌
Co-designer：Chen Long, Xu Tong, Sun Chengwei, Shen Ming, Zheng Bingdong, Xia Lu, Wang Shuyue, Zhao Weilu, Shi Fei, Li Hongpeng, Wang Kai, Liu Che, Wang Pei, Zhou Baochang

非主基地航空公司生活服务设施项目位于大兴国际机场辅助区东北侧,延续一期宿舍(生活服务设施)"户户有阳光"的理念,将宿舍组群呈E形组合,在整体规划中,通过"共享能量树"将两侧E形宿舍区连接起来,有效满足了宿舍区服务半径的要求,以达到功能相对独立、群组形象又完整统一的有机整体。设计积极回应场域环境,利用围合出的内庭院,以及屋顶花园、空中观景台,为入住其中的年轻用户打造丰富、立体的活动休闲空间。对外,建筑保持完整的展示界面以满足机场辅助区城市设计的要求,对内强调建筑界面的变化。在实现户户满足日照要求的同时,将东西向户型外窗做折窗处理,有效地解决了东西晒问题,并使建筑呈现出极为丰富的光影效果,使建筑的尺度更加富有层次。公共服务空间以开放、共享和高效原则集中设置,联系便捷。在北京大兴国际机场非主基地生活服务设施的这一设计实践中,试图通过建筑设计去激发空港社区的活力,最终提升空港员工的生活品质。

The Non-Main Base Airline Living Service Facilities Project is located in the northeast corner of the auxiliary area of Daxing International Airport, continuing the concept of "sunshine for every household" from the first phase of dormitories (living service facilities) . The dormitory clusters are arranged in an "E" shape. In the overall planning, the two sides of the "E" shaped dormitory areas are linked together through "shared energy trees" to effectively meet the service radius requirements of the dormitory area, achieving a relatively independent function, and a unified organic whole with complete and unified group imagery. The design actively responds to the site environment, using enclosed inner courtyards, as well as rooftop gardens and sky observation decks, to create rich and three-dimensional activity and leisure spaces for the young users residing within. Externally, the buildings maintain complete display interfaces to meet the urban design requirements of the airport auxiliary area, while internally emphasizing variations in the building interfaces. While meeting the sunlight requirements of each household, the east-facing and west-facing windows of the unit types are treated with folding windows to effectively solve the problem of sun exposure from the east and west, creating a highly rich effect of light and shadow and enriching the scale of the buildings. Public service spaces are centrally located with principles of openness, sharing, and efficiency, ensuring convenient connectivity. The design practice of the Non-Main Base Airline Living Service Facilities Project at Beijing Daxing International Airport attempts to stimulate the vitality of the airport community through architectural design, and ultimately enhance the quality of life for airport employees.

生长与蔓延 蔓·设计实践 I

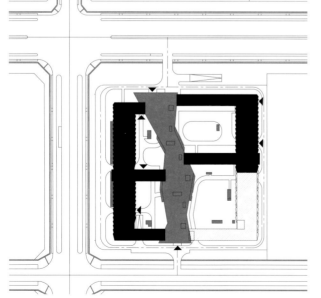

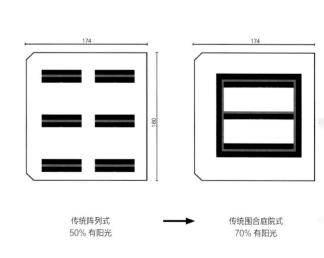

传统阵列式　　　→　　　传统围合庭院式
50% 有阳光　　　　　　　70% 有阳光

总平面 / General Layout
阵列式布局和合院式布局对比 / Comparison between Array Layout and Courtyard Layout

224

北京大兴国际机场非主基地航空公司生活服务设施工程

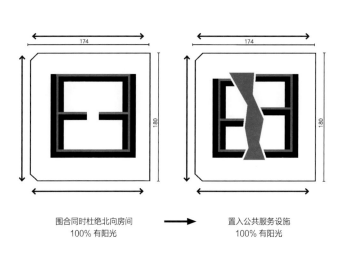

围合同时杜绝北向房间
100% 有阳光

→

置入公共服务设施
100% 有阳光

模型 / Model

北京大兴国际机场非主基地航空公司生活服务设施工程

烟台经济技术开发区景观卫生间项目·Yantai Economic and Technological Development Zone Landscape Toilet Project

设计 Design 2018 · 竣工 Completion 2020

地点：山东，烟台 · 建筑面积：236 平方米
Location：Yantai, Shandong · Floor Area：236m²

合作设计师：李达，郑秉东，陈珑，赵玮璐，邵宇，邓涛，时菲
Co-designer：Li Da, Zheng Bingdong, Chen Long, Zhao Weilu, Shao Yu, Deng Tao, Shi Fei

本项目位于烟台经济技术开发区金沙江路景观带上。主要服务功能包括公共卫生间和公共休息室，设计以"私密和开放"的概念，通过中间走廊将两个功能进行分区，走廊与景观绿化之间的慢行步道形成视觉通廊。卫生间部分采用观感相对柔和的金属穿孔网板作为外饰面材料增强其私密性，休息室部分采用通透的落地玻璃，强调与周边环境的互动联系。整体空间形体通过"抬升"手法，扩大休息室景观面，最大化利用景观，形成"景观取景器"，而合理下压卫生间区域空间以强调其私密感。玻璃、穿孔铝板不同通透感的材料使这一小尺度的建筑呈现出了丰富的层次感。木纹板吊顶与银灰色金属板形成对比，更加增进了建筑的亲切感；外围景观平台做悬挑处理，仿佛一座轻盈漂浮在草坪之上的"城市雕塑"。

This project is located on the landscape belt of Jinshajiang Road in Yantai Economic and Technological Development Zone. The main service functions include public restrooms and public lounges, designed with the concept of "private and open". The two functions are divided by a middle corridor, and a slow walking walkway between the corridor and landscape greening forms a visual corridor. The bathroom adopts a relatively soft metal perforated mesh panel as the exterior decoration to enhance its privacy, while the rest room adopts transparent floor to ceiling glass, emphasizing interaction and connection with the surrounding environment. The overall spatial form is elevated to expand the landscape of the lounge, maximize the utilization of the landscape, and form a "landscape viewfinder". At the same time, the sanitary condition areas are appropriately lowered to emphasize their sense of privacy. The different transparency materials of glass and perforated aluminum panels give this small-scale building a rich sense of hierarchy. The contrast between the wood grain board ceiling and the silver gray metal plate further enhances the building's sense of familiarity. The peripheral landscape platform is cantilevered, like a light floating "urban sculpture" on the lawn.

生长与蔓延　蔓·设计实践 I

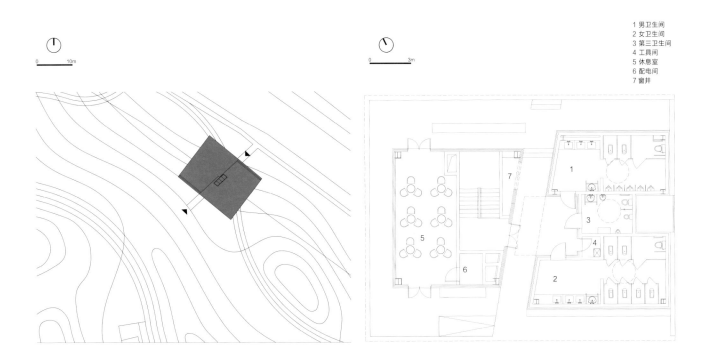

1 男卫生间
2 女卫生间
3 第三卫生间
4 工具间
5 休息室
6 配电间
7 窗井

总平面 / General Layout
首层平面 / First Floor Plan

烟台经济技术开发区景观卫生间项目

烟台蓬莱国际机场信息中心·Yantai Penglai International Airport Information Center

设计 Design 2018 · 竣工 Completion 2023

地点：山东，烟台 · 用地面积：9879 平方米 · 建筑面积：6898 平方米
Location：Yantai, Shandong · Site Area：9,879m² · Floor Area：6,898m²

合作设计师：郑秉东、张良钊、孙成伟、陈珑、徐彤、牟冰峰、邓涛
Co-designer：Zheng Bingdong, Zhang Liangzhao, Sun Chengwei, Chen Long, Xu Tong, Mu Bingfeng, Deng Tao

机场信息中心承担着智慧机场核心系统的计算、存储和信息交互处理的任务，是整个机场的神经中心和"最强大脑"，保证信息安全是实现机场平稳运行的关键所在，本案以"信息·基石"作为设计理念，以坚实、厚重的建筑形态，传达了信息中心应具有的安全可靠和具有科技感的价值取向。建筑设计遵循内容与形式统一的原则，将开窗需求较弱的指挥中心体量与办公功能做虚实处理，并结合建筑朝向，将玻璃幕墙做成锯齿形状，为办公功能争取到最大化的南面朝向，而面向机场的方向，建筑开窗谨慎，以减少噪声。悬挑的指挥中心实体墙面采用人造板材，辅以数字化点阵形式的开窗，以彰显信息建筑的特征。整个建筑强调秩序感和力量感，以期成为烟台机场安全运行的守护者。

The airport information center, responsible for the computation, storage, and processing of information exchange of the intelligent airport core system, serves as the nerve center and "most powerful brain" of the entire airport. Ensuring information security is crucial to achieving smooth airport operations. With the design concept of "Information Keystone", this project emphasizes the solid and substantial architectural form to convey the values of security, reliability, and technological sophistication that an information center should possess. The architectural design follows the principle of unity of content and form, treating the command center volume with weaker window requirements and office functions with a virtual-real approach. Combined with the building orientation, the glass curtain wall is treated in a zigzag pattern to maximize the southern exposure for office functions. Facing the airport, the windows are design with caution to minimize noise. The cantilevered command center uses artificial paneling with digital matrix-style windows to highlight the characteristics of information architecture. The entire building emphasizes a sense of order and strength, aiming to be the guardian of safe operations at Yantai Airport.

生长与蔓延　蔓·设计实践 I

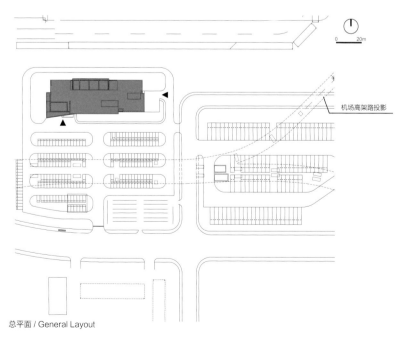

总平面 / General Layout

烟台蓬莱国际机场信息中心

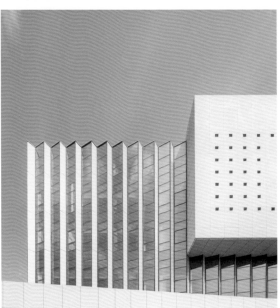

烟台工程职业技术学院南校区工程·South Campus Project of Yantai Engineering & Technology College

设计 Design 2019 · 竣工 Completion 2022

地点：山东，烟台·用地面积：7.34公顷·建筑面积：12.19万平方米
Location：Yantai, Shandong · Site Area：7.34ha · Floor Area：121,900m²

合作设计师：陈珑、李达、徐彤、鲍亦林、张良钊、郑秉东、王舒越、牟冰峰、邵宇
Co-designer：Chen Long, Li Da, Xu Tong, Bao Yilin, Zhang Liangzhao, Zheng Bingdong, Wang Shuyue, Mu Bingfeng, Shao Yu

　　学校是一个知识交流的场所，是技术培训和人才孵化的平台。根据学校产学研发展的规划，将园区围绕运动场呈 U 形布局，采用"筑台"的设计理念，将教学、实训、创业孵化、科研交流等各个功能平台，整合在一起。教学培训区及学生生活区根据地形高差集中布置在场地西南侧，将对外需求较高的实训楼沿昆仑路展开，将具有对外功能的学术交流中心置于场地的东北角，与城市空间充分融合。

　　主教学楼呈 π 字形向城市打开，室外空间结合学生行为模式以及安全疏散等需求，形成立体化的入口广场，广场呈敞开、拥抱之势，增强建筑群体进深感。建筑平面形式高度匹配教学需求，利用地形高差将住宿、餐饮、体育休闲等空间立体化地有机组合，营造出富有活力的课外活动空间。实训楼、学术交流中心依据功能特性，建筑呈开放状态，局部架空处理将城市空间引入体育场，学术交流中心按具备学术研讨接待及城市商务酒店功能的综合体设计。建筑材料采用暖色干挂陶板，与灰蓝色玻璃幕墙组合，以期打造温馨质朴，富有人文及科学精神的校园建筑。

Schools serve as a locus for the exchange of knowledge, as well as a platform for technological training and talent incubation. In alignment with the school's strategic plan for integration of industry, education, and research development, the campus layout adopts a U-shaped arrangement around the sports field, guided by the design philosophy of "platform construction". This concept consolidates various functional platforms—such as teaching, practical training, entrepreneurial incubation, and scientific research—into a cohesive unit. The teaching and training areas, along with student living spaces, are strategically positioned at the southwestern part of the site, taking advantage of the terrain's elevation differences. The practical training building, which has higher external engagement, unfolds along Kunlun Road, while the academic exchange center, with its outward-facing functions, is placed at the northeast corner of the site, fully integrating with the urban space.

The main teaching building opens toward the city in a π-shape, and the outdoor spaces are designed to align with student behavioral patterns and safety evacuation requirements, creating a multi-dimensional entrance plaza. The plaza is open and welcoming, enhancing the depth perception of the architectural ensemble. The architectural layout is highly tailored to educational needs, utilizing the terrain's elevation differences to organically integrate accommodation, dining, and sports and leisure facilities, thereby fostering a vibrant extracurricular activity space. The practical training building and the academic exchange center are designed with an open building to allow the urban space to merge into the sports arena. The academic exchange center is designed as a complex that accommodates academic seminars and city business hotel functions. Building materials include warm-toned dry-hung ceramic tiles combined with grey-blue glass curtain walls, aiming to create a campus building that is both warm and modest, enriched with a spirit of humanity and science.

烟台工程职业技术学院南校区工程

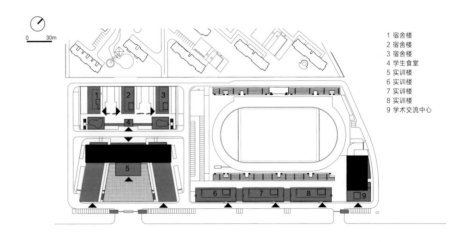

1 宿舍楼
2 宿舍楼
3 宿舍楼
4 学生食堂
5 实训楼
6 实训楼
7 实训楼
8 实训楼
9 学术交流中心

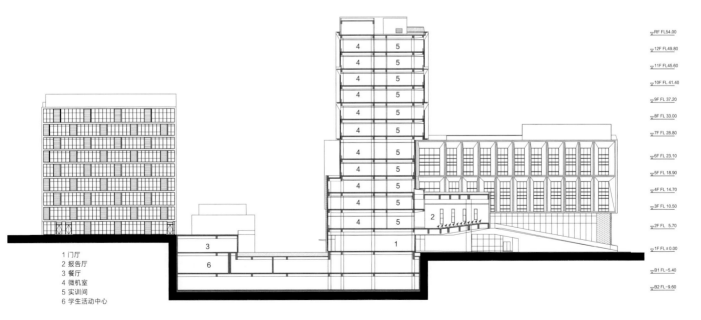

1 门厅
2 报告厅
3 餐厅
4 微机室
5 实训间
6 学生活动中心

剖面 / Section
总平面 / General Layout

烟台中科先进材料与绿色制造山东省实验室·Yantai Zhongke Advanced Materials and Green Manufacturing Shandong Provincial Laboratory

设计 Design 2020·竣工 Completion 2023

地点：山东，烟台·用地面积：21.65 公顷·建筑面积：56.48 万平方米
Location：Yantai，Shandong·Site Area：21.65ha·Floor Area：564,800m²

合作设计师：李达、孙成伟、郑秉东、鲍亦林、许国玺、何霈、徐彤、牟冰峰
Co-designer：Li Da，Sun Chengwei，Zheng Bingdong，Bao Yilin，Xu Guoxi，He Pei，Xu Tong，Mu Bingfeng

本项目的设计区别于传统的"单一性"实验室园区，规划采用核心景观带加环形车道的布局手法，实现了在公园中工作和生活的核心理念。面向城市，建筑呈现科技建筑的理性、简洁和典雅的气质，展现企业形象。面向内部核心景观带，建筑呈现放松、开放状态，强调与景观融合。由外向内逐渐减小的建筑体量，为科研人员营造尺度宜人的内部庭院，缓解工作强度和压力，激活想象力和创造力。主楼由弧形+板形体量组合而成，在入口轴线上架空形成第一层次入口空间，配以景观手法与其他实验室建筑一起，强调了园区的主轴线。实验室功能设计借鉴了国际先进布局模式，充分适应产学研建筑的需求。建筑外立面以干挂陶板为主，强调了科研建筑的性格特征。

Differentiating from traditional "singular" laboratory campuses, the planning utilizes a layout of a core landscape belt with a circular driveway to realize the core concept of working and living in the park. Facing the city, the building exhibits the rationality、simplicity and elegant characteristic of technological buildings, harmonizing with the urban context and showcasing the corporate image. Towards the internal core landscape belt, the building presents a relaxed and open state, emphasizing integration with the landscape. The gradually diminishing building volumes from exterior to interior create pleasantly scaled internal courtyards for researchers, alleviating work intensity and pressure, and stimulating imagination and creativity. The main building is composed of curved and flat volumes, with a raised first-level entrance space along the entrance axis, along with landscape techniques coalescing with other laboratory buildings, enhancing the park's main axis. The laboratory's functional design draws from international advanced layout patterns, fully adapting to the needs of industry-academia research buildings. The building facade predominantly features dry-hung ceramic panels, accentuating the characteristics of research architecture.

生长与蔓延　蔓·设计实践 I

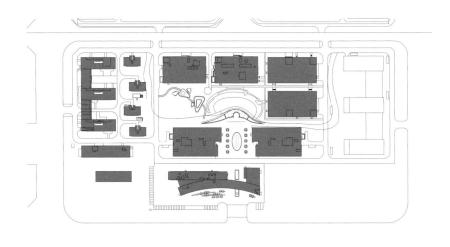

总平面 / General Layout

烟台中科先进材料与绿色制造山东省实验室

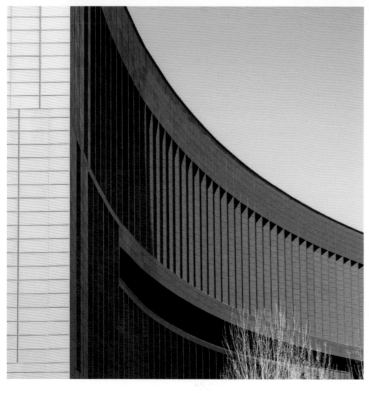

合作设计师：朱谞、孙成伟、丁洋、邵宇、张胜钰、徐彤、李石一
Co-designer: Zhu Xu, Sun Chengwei, Ding Yang, Shao Yu, Zhang Shengyu, Xu Tong, Li Shiyi

烟台光电传感产业园项目·Yantai Photoelectric Sensing Industrial Park Project

设计 Design 2022·竣工 Completion 2024

地点：山东，烟台·用地面积：6.14公顷·建筑面积：8.86万平方米
Location：Yantai, Shandong · Site Area：6.14ha · Floor Area：88,600m²

合作设计师：朱谞、孙成伟、丁洋、邵宇、张胜钰、徐彤、李石一
Co-designer：Zhu Xu, Sun Chengwei, Ding Yang, Shao Yu, Zhang Shengyu, Xu Tong, Li Shiyi

本园区紧贴开发区主干道北京中路，肩负着塑造产业和城市门户形象的重要作用。由于中小体量厂房占比较大，为了形成更加整体统一的园区形象，南北主入口用柱廊雨棚连接两侧建筑，东侧沿北京中路用不同高度的厂房咬合搭接，形成高低起伏、韵律丰富多变的城市天际线，好似一条"光电之链"形成的科技浪潮。园区内部采用人车分流动线，空间结构采用一条景观主轴，生产和生活两核心节点，多个乔木为主、灌木为辅的林下广场，在充满工业元素的氛围中创造近人尺度的中心景观。厂房车间平面规整方正，核心筒及辅助用房布置在建筑两端并采用模块化组合，为生产布局提供了更多灵活性的同时也有效降低了造价。建筑单体用米色国产花岗石为基座，呈托着上部亮银色保温一体铝板外墙，坚实与轻盈的强烈对比，搭配光栅式的竖向富有韵律的开窗形式，营造出坚如磐石根基下的光电科技必将迎来产业的腾飞之势。

Adjacent to the main road of Beijing Middle Road in the development zone, this park plays a crucial role in shaping both industrial and urban gateway images. With a large proportion of medium and small-scale factory buildings, columned corridors connect the north and south main entrances with the surrounding buildings to establish a more unified park image. Along the east side facing Beijing Middle Road, factory buildings of varying heights are juxtaposed to create a skyline with undulating heights, rich rhythms, and dynamic changes, resembling a technological wave formed by a "chain of optoelectronics." Internally, the park adopts separate traffic routes for pedestrians and vehicles, with a spatial structure centered around a landscape axis. The production and living core nodes are surrounded by several deciduous trees and shrubs, creating central landscapes on a human scale within an industrial atmosphere. The layout of the factory workshops is regular and square, with core tube and ancillary buildings arranged at both ends of the buildings in a modular combination, providing greater flexibility in production layout while effectively reducing costs. The individual buildings are based on beige domestic granite, supporting upper parts clad in bright silver integrated aluminum panels, creating a strong contrast between solidity and lightness, complemented by a rhythmic vertical window pattern, evoking an image of optoelectronic technology poised for industrial takeoff on a rock-solid foundation.

生长与蔓延 蔓·设计实践 I

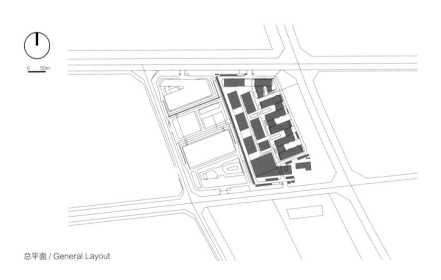

总平面 / General Layout

烟台光电传感产业园项目

北京大兴国际机场行政综合业务楼 · Beijing Daxing International Airport Administrative Comprehensive Business Building

设计 Design 2016 · 竣工 Completion 2024

地点：北京 · 用地面积：2.48 公顷 · 建筑面积：9.30 万平方米
Location：Beijing · Site Area：2.48ha · Floor Area：93,000m²

合作设计师：郑秉东、李达、夏璐、陈珑、孙成伟、徐彤、朱谞、郭锦洋、李石一、崔伟、邓涛、时菲
Co-designer：Zheng Bingdong, Li Da, Xia Lu, Chen Long, Sun Chengwei, Xu Tong, Zhu Xu, Guo Jinyang, Li Shiyi, Cui Wei, Deng Tao, Shi Fei

本项目位于北京大兴国际机场出场高架的起始位置上，是旅客往来机场必经的一栋建筑。设计以"北京门户"的概念为依据来呼应其所处的显要位置和重要功能。行政综合业务用房的两个办公单位呈U形组合，相对独立又在建筑形象上相互连接，与地块西侧待建的业务办公衔接围合成有机的整体；同时，提取出建筑群中展示、服务等公共性功能，采用体块介入、虚实对比的造型手法，在规则合院造型中斜向嵌入具有视觉冲击力的异形玻璃体量，形成建筑焦点，以轻巧、通透而又坚实有力的形象营造出一种崛起之势；围合的院落使建筑最大限度地拥抱景观环境，斜插的通透玻璃体也起到进一步放大和激活景观空间的效果；建筑主体表皮选择素雅大气的干挂陶板，与体现现代科技感的菱形玻璃幕墙相得益彰，展现出大兴国际机场与时俱进的前瞻性设计；综合管理、业务办公楼与工程档案、展示馆和服务中心多种功能呼应融合，互补共生，成为新机场的重要展示窗口。

As the project is located at the staring point of the exit flyover from Beijing Daxing International Airport, it is a building that passengers inevitably pass by while arriving at and departing from the airport. Based on the concept of "Beijing Gateway", the design highlights its prominent position and crucial functions. The two office blocks of the administrative integrated business building, arranged in a U-shaped combination, maintain relative independence while being mutually connected, forming an organic whole with the business office to be built on the west side of the plot. At the same time, key public functions such as exhibition and services are extracted from the building complex and integrated into the design using block insertion and a solid-void contrast approach. In the regular courtyard layout, a diagonally inserted, uniquely shaped glass volume with strong visual impact becomes the architectural focal point. This composition evokes a sense of rise through its light, transparent, yet solid and forceful imagery. The enclosed courtyard maximizes the building's connection to the surrounding landscape, while the transparent, slanted glass volume further amplifies and activates the outdoor space. The building's exterior is clad in elegant, dry-hanging ceramic panels, which harmonizes with the modern, futuristic feel of the uniquely shaped glass curtain wall, reflecting the forward-thinking design of Beijing Daxing International Airport. The integrated management and business office building echoes the multiple functions of the project archive and exhibition pavilion, and service center, complementing each other to become an important display window for the new airport.

生长与蔓延　蔓·设计实践 I

北京大兴国际机场行政综合业务楼

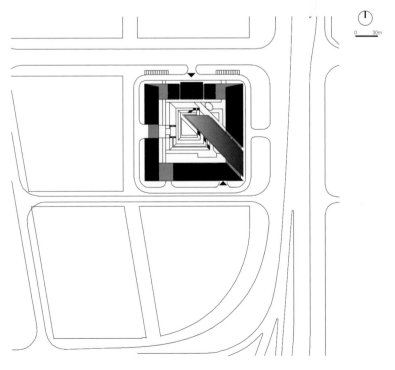

总平面 / General Layout

紫光南京集成电路基地项目（一期）· Tsinghua Unigroup Nanjing Integrated Circuit Base Project Phase I

设计 Design 2017

地点：江苏，南京 · 用地面积：39.59 公顷 · 建筑面积：55.19 万平方米
Location: Nanjing, Jiangsu · Site Area: 39.59ha · Floor Area: 551,900m²

合作设计师：李达、郑秉东、陈珑、孙成伟、夏璐、鲍亦林、赵玮璐
Co-designer: Li Da, Zheng Bingdong, Chen Long, Sun Chengwei, Xia Lu, Bao Yilin, Zhao Weilu

　　本项目在尊重芯片生产工艺及动力传输、材料运输等各种需求的前提下，强调现代高科技厂区高效便捷智能的同时，更加注重通过丰富的高品质室内外工作空间的打造，来激发员工交流交往的积极性，从而提高研发人员的工作主动性、创造力。总体布局通过对工艺需求的研究，将厂区各功能区呈平行化布置，然后用最短的连廊将其串联起来，最后由厂区的弹性空间将整个建筑群整合在一起。FAB核心厂房包括设备机房层、技术夹层和洁净生产层，两端为辅助支持区，内部功能自然形成一个U形的剖面结构，U形两端高出的女儿墙为屋面设备用房。厂区建筑的设计提取核心厂房的U形形态作为原型。通过系统的建筑语言，以U形为基本单元，根据不同厂房的功能形成几个大小体量的U形组合，通过厂区弹性空间整合在一起。建筑群面向景观河道形成四百多米长的建筑立面，高低起伏，生动有序。有效地将超大的工业建筑尺度消解平衡为城市尺度。根据各栋建筑的功能特性赋予不同通透性的立面材料。运维楼以玻璃幕墙与点窗组合，实现了良好的采光效果和景观视野。厂房立面开窗较少，通过铝板，穿孔板，波纹板等材料的组合搭配，形成统一而丰富的立面效果。研发楼更是结合竖向上不同功能需求，分别赋予从开放、封闭再到开放的不同的建筑表情。其他辅助建筑同样以"形式追随性能"的原则，投入同等的关注，整个厂区力图以强烈的设计语言彰显现代高科技工业建筑的美学特征和品质。

This project prioritizes efficiency, convenience, and intelligence within a modern high-tech industrial park environment, while accommodating various demands such as chip manufacturing processes, power transmission, and material transportation. It places significant emphasis on enhancing the quality of both indoor and outdoor work spaces to encourage active communication and interaction among employees, thereby fostering initiative and creativity among research and development personnel. The overall layout is developed based on an analysis of process requirements, with functional areas of the plant arranged in a parallel configuration and interconnected by the shortest possible corridors. The architectural complex is unified through flexible spaces within the plant. The core FAB building features layers for equipment machinery, a technical interlayer, and a clean production area, flanked by support zones at both ends, forming a U-shaped sectional structure. Parapet walls at the ends of the U-shape accommodate rooftop equipment. The architectural design of the plant extracts the U-shaped form of the core building as a prototype, with multiple U-shaped combinations of varying sizes formed according to the functions of different buildings. These combinations are integrated through the plant's flexible spaces. Facing the landscape river, the architectural complex forms a facade over four hundred meters long, undulating and orderly, effectively reducing the vast industrial scale to an urban scale. Facade materials with varying degrees of transparency are assigned based on the functional characteristics of each building. The maintenance building combines glass curtain walls and punctuated windows, achieving excellent lighting effects and scenic views. The plant's facade, with limited window openings, utilizes a combination of aluminum panels, perforated plates, and corrugated panels to create a unified and rich facade effect. The research and development building integrates vertical functionality requirements, providing expressions from open to closed and back to open. Other auxiliary buildings also adhere to the principle of "form follows function" and have received significant attention. The entire factory area aims to showcase the aesthetic characteristics and quality of modern high-tech industrial buildings with a robust design language.

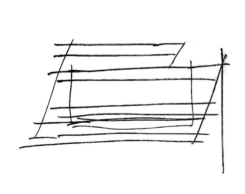
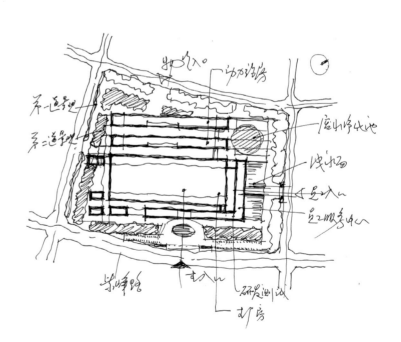

设计草图 / Sketch

厂区建筑原型提取－主厂房 U 形剖面

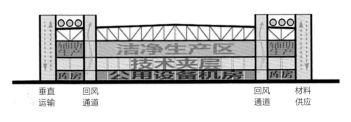

总平面 / General Layout
剖面 / Section

紫光南京集成电路基地项目（一期）

生长与蔓延　蔓·设计实践 I

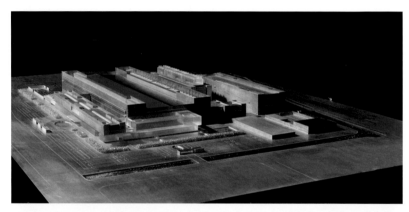

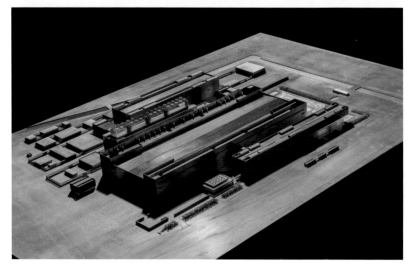

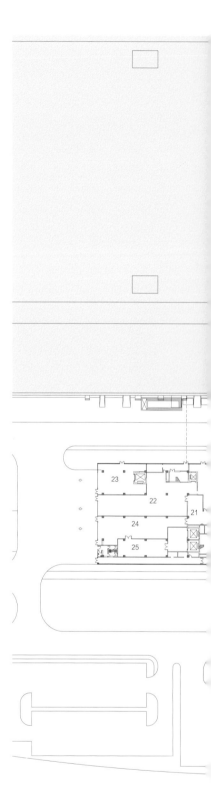

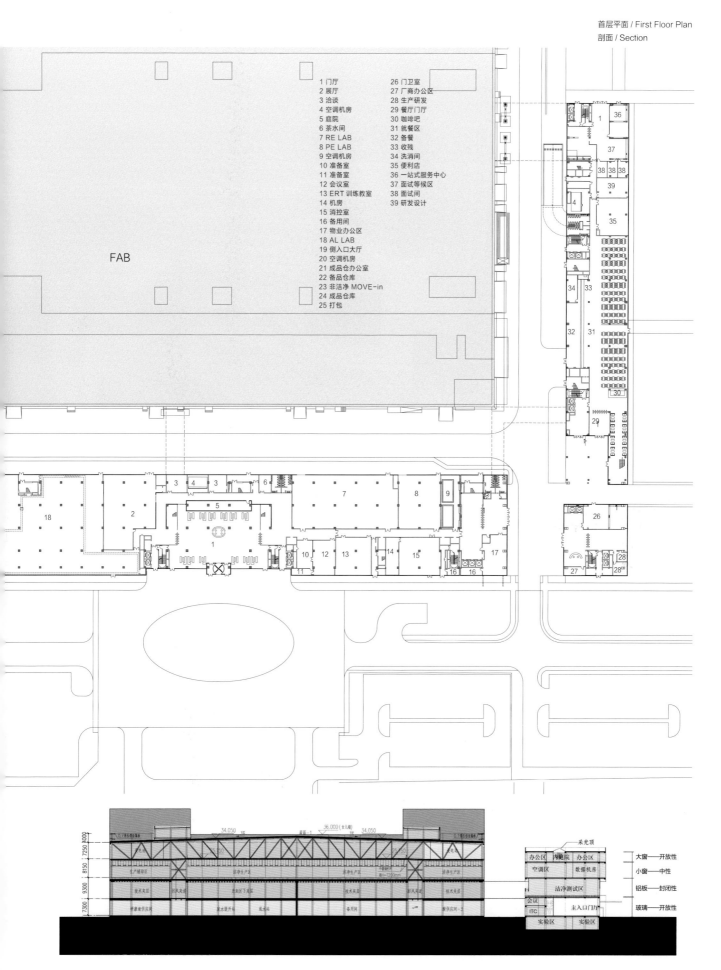

潍坊云创金谷·Weifang Yunchuang Golden Valley

设计 Design 2022

地点：山东，潍坊 · 用地面积：5.68 公顷 · 建筑面积：23.20 万平方米
Location：Weifang, Shandong · Site Area：5.68ha · Floor Area：232,000m^2

合作设计师：夏璐、陈珑、孙成伟、张良钊、鲍亦林、徐彤、邵宇、朱谓、李天宇、许国玺、牟冰峰、郭锦洋、赵玮璐、何霈、张胜钰、王舒越、邓涛、时菲
Co-designer: Xia Lu, Chen Long, Sun Chengwei, Zhang Liangzhao, Bao Yilin, Xu Tong, Shao Yu, Zhu Xu, Li Tianyu, Xu Guoxi, Mu Bingfeng, Guo Jinyang, Zhao Weilu, He Pei, Zhang Shengyu, Wang Shuyue, Deng Tao, Shi Fei

该建筑为典型的超大型城市综合体，面向城市主干道，以"云平台"的概念，将购物、娱乐、餐饮、酒店、公寓等功能整合在一起，展现出大气、多元、浪漫的气息，在城市主要道路方向力图塑造一种"城市力量"，她是一个承载着科技、文化、自然、商业、娱乐的平台，并且将其底部活跃、跳动的建筑体量整合统一；场地东南侧，在开敞的下沉广场前，五层裙楼通过退台处理，形成餐饮外摆区的同时，营造自由律动的"云谷空间"，充分利用自然阳光，与下沉广场形成对话，将内部庭院空间彻底激活。从下沉广场、天光中庭、屋顶花园到24小时酒吧街，多样的特色场所营造了丰富的空间体验，使顾客充分享受现代城市商业空间给人带来的幸福体验感。

This building typifies a large-scale urban complex, facing the city's main thoroughfare. Conceptualized around the idea of a "cloud platform", it integrates functions such as shopping, entertainment, dining, hotels, and apartments, exuding an atmospheric, diverse, and romantic ambiance. Aimed at shaping a sense of "urban vitality" along the city's major road, it serves as a platform that integrates technology, culture, nature, commerce, and entertainment, harmonizing its active and pulsating architectural volumes at the base. On the southeastern side of the site, in front of the open sunken plaza, the five-story podium is designed with a stepped-back treatment, creating an outdoor dining area while shaping a freely flowing "Cloud Valley Space". This design maximizes the use of natural sunlight and establishes a dialogue with the sunken plaza, thoroughly enlivening the internal courtyard space. From the sunken plaza and sky-lit atrium to rooftop gardens and a 24-hour bar street, a variety of distinctive spaces offer a rich spatial experience, allowing customers to fully enjoy the happiness brought by modern urban commercial spaces.

生长与蔓延 蔓·设计实践 I

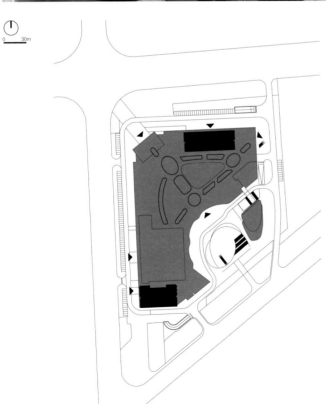

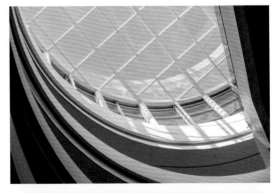

总平面 / General Layout
模型 / Model

潍坊云创金谷

珠海机场新建塔台及配套项目·Zhuhai Airport New Air Traffic Control Tower and Supporting Building Project

设计 Design 2022

地点：广东，珠海 ·用地面积：1.37 公顷 ·建筑面积：1.82 万平方米
Location：Zhuhai, Guangdong · Site Area：1.37ha · Floor Area：18,200m²

合作设计师：孙成伟、郑秉东、鲍亦林、邵宇、徐彤、姚二将、邓涛、时菲
Co-designer: Sun Chengwei, Zheng Bingdong, Bao Yilin, Shao Yu, Xu Tong, Yao Erjiang, Deng Tao, Shi Fei

项目位于中国典型滨海城市——珠海。由于塔台和空管楼面向机场主跑道并紧邻航站楼，故设计注重将建筑与机场整体统一考虑。该项目引入"空中冲浪"的设计理念，塔台利用海浪的造型意向，回应珠海的自然环境、城市气质与文化内涵。该项目建筑群为设备用房提供充足、完整、高效空间的同时，力求让航管人员在自然中工作和生活。通过精心的设计，原本零散的场地被整合成尺度宜人的花园景观与共享绿色庭院，以缓解航管人员的工作强度和压力，提高工作效率。项目的长远目标是塑造标志性的塔台形象、高效运行的设备系统、人与自然融合的复合型空间环境。塔与海融合的设计概念，最终旨在将其打造成为具有标志性、高效性、可持续性，并达到世界一流水平的滨海机场塔台及空管园区。

The project is located in Zhuhai, a typical coastal city in China. Since the control tower and the air traffic control building face the main runway of the airport and their proximity to the terminal building, the design emphasizes a harmonious integration between the buildings and the overall image of the airport, as well as the character of the city of Zhuhai. Incorporating the concept of "surfing in the sky", the tower's design employs the imagery of ocean waves, echoing Zhuhai's natural environment, urban temperament, and cultural depth. The project's architectural complex provides ample, comprehensive, and efficient spaces for equipment rooms while striving to enable air traffic controllers to work and live amidst nature. Through meticulous planning, the originally fragmented site has been transformed into a series of well-proportioned garden landscapes and shared green courtyards, aiming to alleviate the workload and stress of air traffic controllers, thereby enhancing their work efficiency. The long-term vision of the project is to create a landmark tower image, equipped with a highly efficient operational system, fostering a composite spatial environment where humans and nature coexist harmoniously. The design concept of "the integration of tower and sea" ultimately aims to establish a tower and air traffic control park that is not only iconic, efficient, and sustainable but also meets world-class standards.

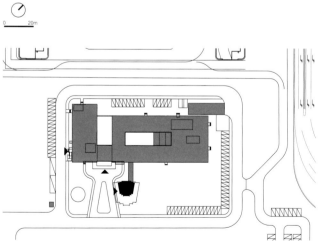

总平面 / General Layout

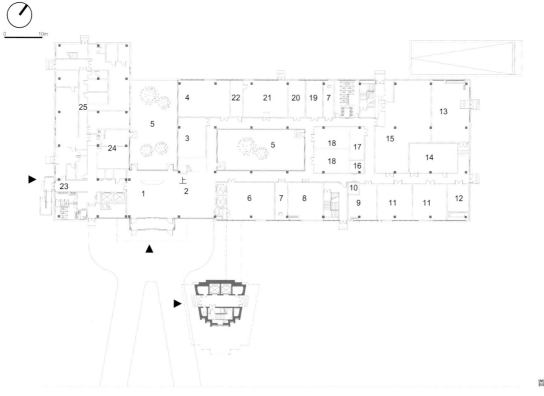

1 门厅
2 电梯厅
3 展厅
4 接待室
5 景观庭院
6 航展飞行讲解室
7 进线间
8 弱电机房
9 电池室
10 弱电间
11 UPS间
12 电池室
13 柴发室
14 高压配电
15 低压配电
16 强电间
17 气灭钢瓶间
18 备件间
19 后勤维修
20 库房
21 消防、安防控制室
22 值班室
23 宿舍门厅
24 业务洽谈室
25 厨房区

首层平面 / First Floor Plan

厦门天马光电子有限公司第 8.6 代新型显示面板生产线项目·Xiamen Tianma Optoelectronics Co., Ltd 8.6 Generation New Display Panel Production Line Project

设计 Design 2022

地点：福建，厦门 · 用地面积：64.37 公顷 · 建筑面积：94.70 万平方米
Location：Xiamen, Fujian · Site Area：64.37ha · Floor Area：947,000m²

合作设计师：夏璐、郑秉东、孙成伟、郭锦洋、李天宇
Co-designer: Xia Lu, Zheng Bingdong, Sun Chengwei, Guo Jinyang, Li Tianyu

本项目总体设计以"工艺路径最短化"为原则，在总体布局上将超大尺度的生产空间后撤，让出南广场，依次布置厂前区、核心生产区、辅助支持区与预留区，厂前区将研发、展示、会议交流和员工宿舍根据功能需求组合在一起。研发办公楼、餐厅和报告厅等功能围合出U形建筑体量，环绕入口礼仪广场，依托尺度庞大厚重的核心生产区厂房，研发管理楼则通过简洁的整体弧线造型，一方面实现与天马老厂建筑文脉的延续，一方面将厂前区、后厂房有机联系起来，打造刚柔相济的总体形态和多层次的厂区空间。超长水平延展的弧形体量，结合CI系统认知，塑造具有天马企业极强的场域感和建筑识别性。辅助区和生产区一体化处理，在满足核心厂房因洁净要求尽量少开窗的前提下，结合通风百叶、消防救援、设备搬运口等增强尺度感和构成感，强调不同尺度的整合性。研发管理区外立面玻璃幕墙上的V形水平金属百叶的设计，呼应"天马展翅"的企业理念。生活区的设计将居住建筑公建化，使宿舍区与厂前区建筑有机的结合在一起。

The primary design principle of this project is "to streamline the process pathway". In the overall layout, the ultra large scale production space will be arranged to the back, resulting in the creation of the southern plaza. The layout comprises the front factory area, the core production zone, the auxiliary support area, and the reserved zone, arranged sequentially. The front factory area accommodates various functions such as research, display, conferences, and employee dormitories, tailored to functional needs. A "U"-shaped architectural volume, comprising the research office building, restaurant, and auditorium, encloses an entrance ceremonial square. Building upon the substantial scale of the core production zone factory building, the research management building adopts a concise arcuate design that harmonizes with the architectural context of the Tianma old factory. This design seamlessly connects the front factory area with the rear factory building, creating a cohesive form and multi-level factory space. The elongated horizontal arcuate volume, combined with the CI system, establishes a distinct architectural identity for Tianma. The auxiliary and production areas are integrated, meeting minimal window requirements for cleanliness in the core factory building while achieving enhanced scale and composition through features like ventilation louvers, fire rescue facilities, and equipment transfer ports. The horizontal metal louvers in the glass curtain wall facade of the research management area reflect the corporate concept of "Tianma spreading its wings." Additionally, the design of the living area transforms residential buildings into public structures, seamlessly integrating the dormitory area with the buildings in the front factory area.

生长与蔓延　蔓·设计实践 I

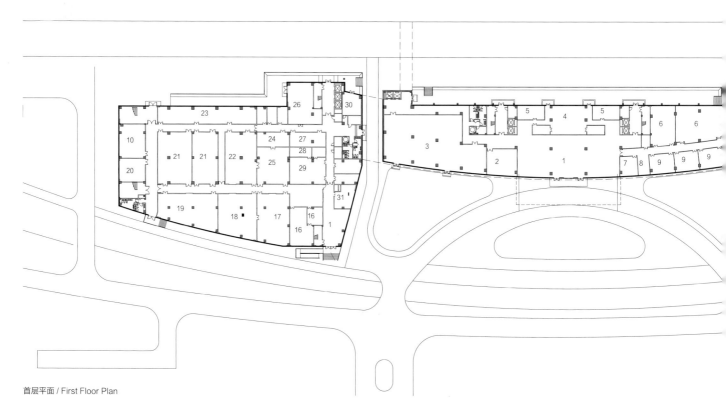

首层平面 / First Floor Plan

厦门天马光电子有限公司第 8.6 代新型显示面板生产线项目

1 门厅
2 VIP 会客厅
3 展厅
4 休息茶座
5 VIP 接待室
6 VIP 会议室
7 行政
8 茶水间
9 会议室
10 空调机房
11 餐厅
12 超市
13 水吧
14 储藏间
15 厨房区
16 办公区
17 电子综合实验室
18 光学综合实验室
19 ACD/OD 实验室
20 业务用房
21 备品间
22 视效检验间
23 电子分析实验室
24 结构分析室
25 力学综合实验室
26 变电所
27 样品间
28 会议室
29 战情室
30 消防设备间
31 接待室

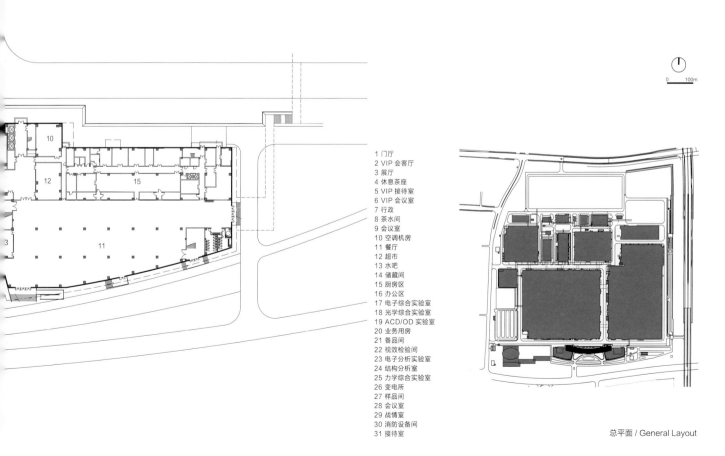

总平面 / General Layout

广州大湾区数字经济和生命科学产业园项目·Guangzhou Greater Bay Area Digital Economy and Life Science Industrial Park Project

设计 Design 2024

地点：广东，广州·用地面积：6.24 公顷·建筑面积：30.46 万平方米
Location：Guangdong, Guangzhou · Site Area：6.24ha · Floor Area：304,600m^2

合作设计师：朱谞、鲍亦林、孙成伟、郑秉东、许国玺、张胜钰、徐彤、丁洋、郭锦洋、刘敬一、卫君
Co-designer: Zhu Xu, Bao Yilin, Sun Chengwei, Zheng Bingdong, Xu Guoxi, Zhang Shengyu, Xu Tong, Ding Yang, Guo Jinyang, Liu Jingyi, Wei Jun

项目紧邻广州南沙明珠湾大桥，快速接驳广佛高速，40min可达珠江新城，快速通达广州市区，地理位置卓越。区别于传统的"单一封闭性"产业园区，在本次设计中提出"南沙之链"的设计概念，将独立分散的两个地块串联起来，成为了城市绿化的延续。在公园中生产和生活，是我们倡导的核心理念。"南沙之链"由三部分构成：城市链——园区为开放型园区，北侧沿街为主要的到达界面，建筑布局退让街角，犹如张开的臂膀拥抱着城市，吸引更多的瞩目与对话。东侧沿街首层为配套用房，建筑首层局部架空二层出挑，为城市居民和园区员工搭建了停留与共享的空间。生态链——规划布局遵循城市主导风向，打造中央庭院，设置建筑架空，保证自然风的贯穿与畅通；采用庭院地下室、立面遮阳、空中绿化、屋面光伏等绿色建筑手法和能源策略。产业链——园区主要功能空间生产—研发—办公—宿舍采用环行布局，员工可以通过内院的空中廊桥和东侧沿街二层的公共服务内街实现无阻碍的跨街连通，便捷的达到园区每一个功能空间，实现生产办公的高效率。

The project is situated next to Mingzhuwan Bridge in Nansha, Guangzhou, providing swift connectivity to the Guangfo Expressway, allowing a 40-minute commute to Zhujiang New Town and efficient access to downtown Guangzhou, making its geographical location outstanding Distinct from the traditional "monolithic and enclosed" industrial park model, this design introduces the "Nansha Chain" design concept, linking two independent and dispersed parcels, thereby extending the urban green space. Our core philosophy advocates for a lifestyle where production and living merge within a park setting. The "Nansha Chain" is comprised of three parts: Urban Chain – an open park along the northern street edge that serves as the primary arrival interface. The building layout is set back from the street corner, resembling open arms that embrace the city, attracting attention and fostering interaction. The first floor along the eastern street is dedicated to ancillary facilities, featuring partially elevated first-floor structures with second-floor overhangs, creating a shared space for city residents and park employees. Ecological Chain – The layout follows the prevailing urban wind direction, creating a central courtyard with elevated structures to ensure natural airflow and circulation. Green building strategies and energy-efficient measures are employed, such as an underground courtyard, sun-shaded facades, rooftop greenery, and photovoltaic panels. Industrial Chain – Key functional areas, including production, R&D, offices, and dormitories, are arranged in a ring formation. Employees can move seamlessly between functional areas through elevated walkways within the courtyard and an interior street on the second floor along the eastern facade, facilitating efficient, cross-functional connectivity across the park.

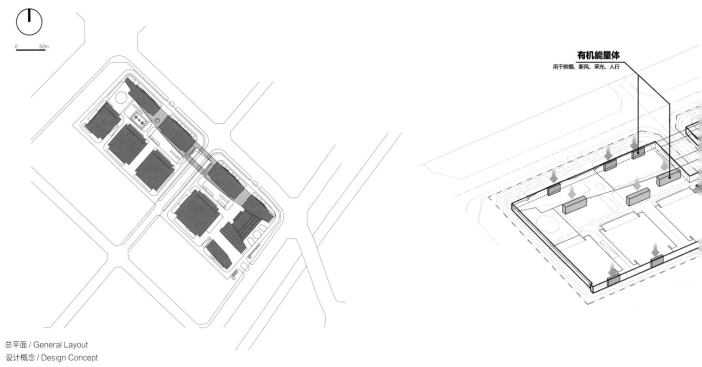

总平面 / General Layout
设计概念 / Design Concept

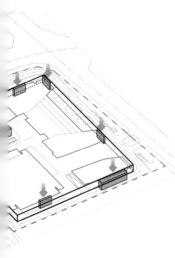

中国国学中心竞赛 · China Sinology Center Competition

设计 Design 2010

地点：北京 · 用地面积：1.62 公顷 · 建筑面积：5.97 万平方米
Location: Beijing · Site Area: 1.62ha · Floor Area: 59,700m²

合作设计师：孙成伟、权薇、董召英、朱谞、邓涛
Co-designer: Sun Chengwei, Quan Wei, Dong Zhaoying, Zhu Xu, Deng Tao

中华民族刚健方正、开放包容，层叠的典籍筑成中华文明之塔，承载历史、立足当代、启迪未来。高耸挺拔的建筑形象寓意着人才辈出的中华儿女，在对国学探索和实践中把中华文明不断带向新的高峰。该项目采用"文明之塔"的设计理念，方案从城市规划和本身的立意选择了用足80m限高的塔形造型——"由层叠的书垒成的文明之塔"，更深层的寓意是"唯有基座之上的九层叠涩塔体才能象征作为泱泱大国固有之学术中心的崇高和博大"。同时，希望位于奥林匹克中心区东北部文化综合区的最南端的国学中心能以较醒目的形态，对其北侧的国家美术馆、工艺美术馆、非物质遗产馆形成统领之势。功能采用竖向分区，下"展"上"研"，展览部分将交通核外置、中心为完整大空间以方便灵活布展，研究功能则以空中四合院的形态置于顶部；古铜色的"多宝格"包裹着充满宝藏的国学中心，并可以起到遮阳作用；外轮廓上扩下收的处理，使整个建筑显得既大气、厚重、挺拔，又蕴有升腾之势。

The robust and upright nature of the Chinese nation, both open and inclusive, constructs a towering edifice of Chinese civilization, built upon layers of timeless classics. It bears the weight of history, stands firmly in the contemporary era, and illuminates the future. The soaring architectural form epitomizes the perpetual emergence of talented Chinese individuals, propelling Chinese civilization to new heights through the exploration and practice of national studies. The project adopts the design concept of "Tower of Civilization", with the plan selecting an imposing 80-meter-tall tower form that is not just an urban landmark but also embodies an intrinsic vision—a tower crafted from stacked volumes of knowledge, where the deeper symbolism lies in the fact that "only the nine-tiered tower above the base can truly represent the grandeur and scope of the academic hub of a great nation". Located at the southernmost tip of the cultural complex, northeast of the Olympic Center, the National Studies Center stands proudly, overseeing its neighboring institutions—the National Art Museum, Craft Art Museum, and Intangible Heritage Museum. Functionally, it is vertically divided, with exhibition spaces occupying the lower levels and research facilities situated at the summit. The exhibition areas feature externalized traffic cores and centralized large spaces, allowing for flexibility in layout, while the research facilities occupy a serene sky courtyard at the top. The building is enveloped by an antique bronze "treasure grid", not only housing the precious contents of the National Studies Center but also serving as a functional sunshade. Its outward expansion and inward tapering silhouette lend the structure a sense of solidity, grandeur, and upward momentum.

生长与蔓延　蔓·设计实践 I

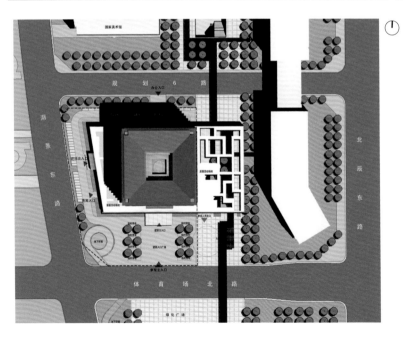

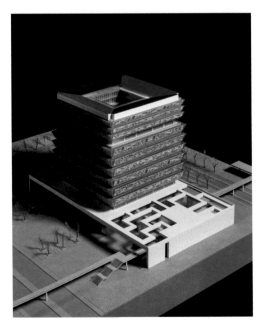

总平面 / General Layout
模型 / Model

中国国学中心竞赛

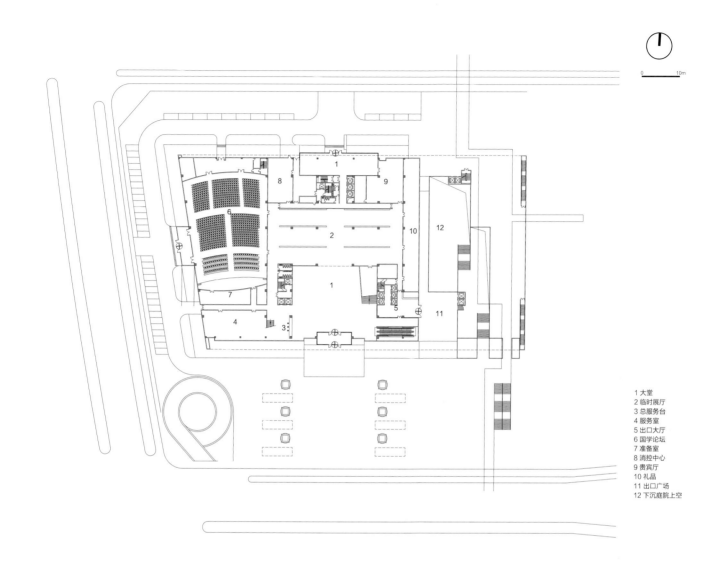

1 大堂
2 临时展厅
3 总服务台
4 服务室
5 出口大厅
6 国学论坛
7 准备室
8 消控中心
9 贵宾厅
10 礼品
11 出口广场
12 下沉庭院上空

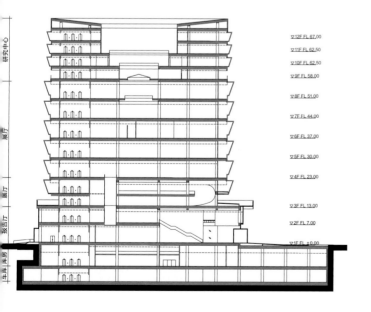

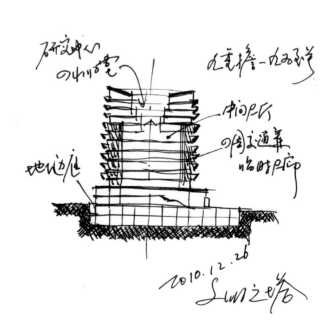

首层平面 / First Floor Plan
剖面 / Section

279

中国瑞达投资发展集团公司瑞达石景山路 23 号院科研基地·China Ruida Investment and Development Group Corporation Ruida Shijingshan Road No. 23 Academy Scientific Research Base

设计 Design 2013

地点：北京 ·用地面积：3.11 公顷 ·建筑面积：14.82 万平方米
Location：Beijing · Site Area：3.11ha · Floor Area：148,200m²

合作设计师：李达、孙成伟、徐彤、邓涛
Co-designer: Li Da, Sun Chengwei, Xu Tong, Deng Tao

本项目南邻长安街，西靠五环，优越的地理位置是打造地标建筑，展示企业形象的最佳之地。因基地西侧和北侧紧邻住宅，在设计中充分考虑新建建筑对住宅的遮挡因素，根据日照计算结果和功能需求，对建筑形体进行消减，同时营造出饱览北京西山风景的景观办公空间。连绵起伏的链状体量通过与沿主要街道的旧办公楼进行连接，在西长安街及建筑东侧创造了生动的立面效果。新建筑的介入极大限度地提升了该地块的用地价值和空间活力。

Located south of Chang'an Avenue and adjacent to the Fifth Ring Road, this site is ideally positioned for creating landmark architecture that showcases corporate image. Given the proximity of residential areas to the west and north of the site, the design considers shading factors for residences, using sunlight analysis and functional demands to reduce the mass of the buildings while creating landscape office spaces that offer views of Beijing's Western Hills. The undulating chain-like mass connects with old office buildings along the main street, creating lively facade effects on West Chang'an Street and the east side of the building. The intervention of new architecture maximally enhances the land value and spatial vitality of the site.

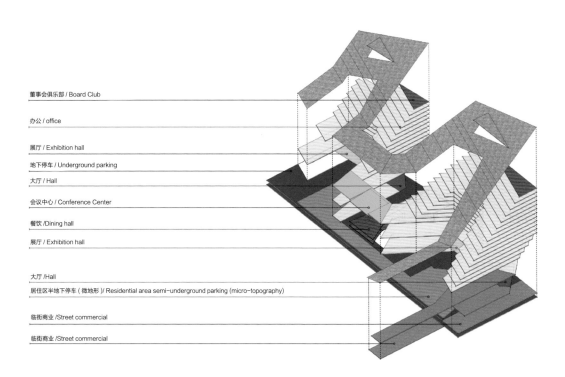

功能分析 / Function Analysis

中国瑞达投资发展集团公司瑞达石景山路 23 号院科研基地

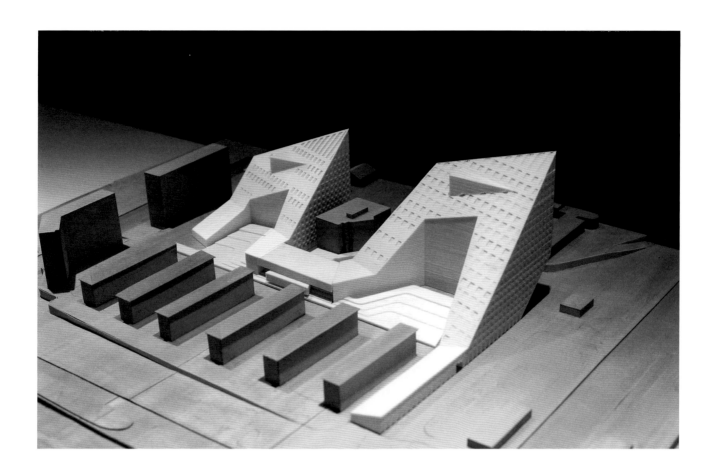

STEP1: 基地范围与保留建筑。

STEP2: 建筑东西侧退让，应对居民区的压力和生成建筑办公入口。

STEP3: 打开，生成庭院，作为居住区与保留建筑的缓冲区域。

STEP4: 建筑对居民区遮挡严重。

STEP5: 建筑西侧压低，东侧提升，利用太阳光角原理，将新建筑对居民区遮挡降到最低。

STEP6: 连接场地西侧保留建筑，拓展长安街立面。

STEP7: 生成建筑住入口。

STEP8: 与景观结合，概念生成结束。

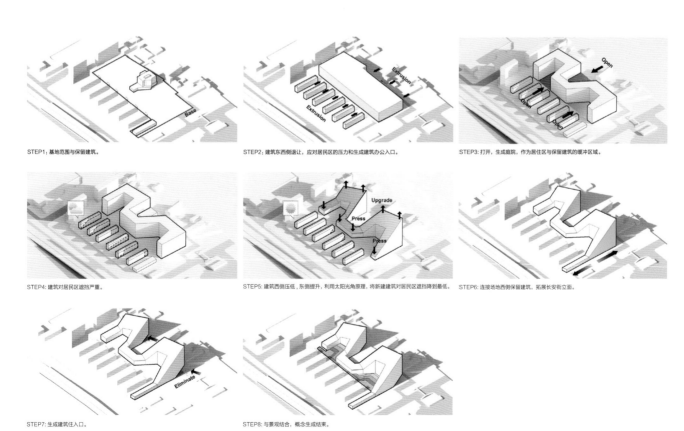

模型 / Model
概念生成 / Concept Generation

中国计算机博物馆项目概念性规划方案设计·Conceptual Planning Scheme Design of the Computer Museum Project

设计 Design 2022

地点：浙江，横店·用地面积：6.67公顷·建筑面积：6.01万平方米
Location: Hengdian, Zhejiang · Site Area: 6.67ha · Floor Area: 60,100m²

合作设计师：夏璐、郑秉东、朱谞、徐彤、邵宇、李天宇、丁洋、郭锦洋、张然然、梁波、姚二将、李石一、张胜钰
Co-designer: Xia Lu, Zheng Bingdong, Zhu Xu, Xu Tong, Shao Yu, Li Tianyu, Ding Yang, Guo Jinyang, Zhang Ranran, Liang Bo, Yao Erjiang, Li Shiyi, Zhang Shengyu

　　人类文明的升级就是依附于人类不断克服一个个认知障碍，计算机的发明使人类来到了信息时代。虽然我们无法预测未来世界将如何发展，但我们可以期待未来的世界必将是人、科技、自然深度融合的状态。因此，我们希望通过"智慧阶梯"的设计理念，由博物馆这一阶梯将人文城市环境与巍巍自然和谐融为一体，最终打造一座具有创新性、前瞻性、可持续性的面向世界的一流高科技博物馆和参与感极强的城市乐园。场地南高北低，同时由于科普馆和未来馆对展示空间的高度需求越来越高，因此建筑从北向南依次排列入口广场、历史馆、科普馆和未来馆，呈现节节升起的态势，从而达到理念、功能与形式的高度统一。从科技到自然，建筑从流动的电路渐变为自然曲线，连续的上人屋面可以作为大小不一的室外科普展场，同时形成一座智慧的阶梯，象征计算的脚步历阶而上，走向未来。建筑外墙结合现代数字显示技术与金属板数字镂空处理，彰显其高科技展示功能。

The evolution of human civilization is fundamentally rooted in the ability to overcome cognitive barriers, and the advent of the computer has ushered humanity into the information age. Although the future development of the world remains unpredictable, we envision a future where humans, technology, and nature are intricately intertwined, fostering a harmonious coexistence that enhances our understanding and experience of the world around us. Thus, guided by the design principle of "Stairs of Wisdom," the museum endeavors to blend the urban human environment with the splendor of nature, endeavoring to construct an innovative, forward-looking, and sustainable high-tech museum and urban park with a strong sense of participation that resonates on a global scale. The site is high in the south and low in the north, facilitating a sequential arrangement of spaces from north to south: the entrance plaza, the History Hall, the Science Hall, and the Future Hall. This layout, resembling an ascending gesture, achieves a cohesive integration of concept, function, and form. Transitioning from technology to nature, the architecture evolves from flowing circuits to organic curves, while continuous accessible rooftops offer diverse outdoor science exhibition spaces, symbolizing the steps of progress towards the future. The building facade incorporates modern digital display technologies and metal plate digital hollowing treatment, accentuating its high-tech exhibition functionality.

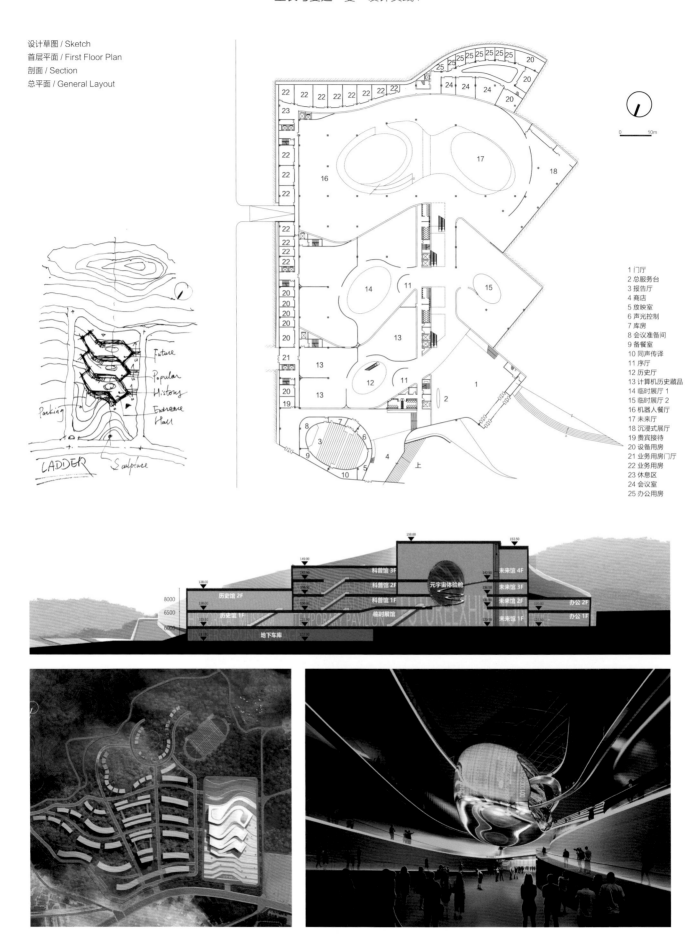

中国计算机博物馆项目概念性规划方案设计

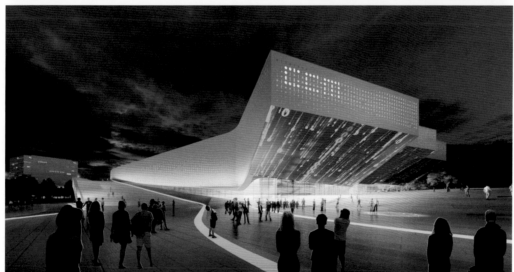

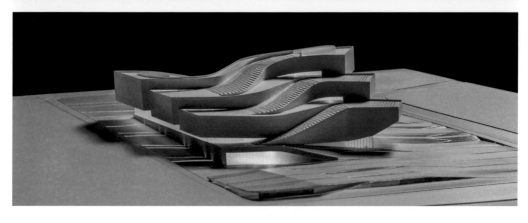

模型 / Model

学术访谈

Academic Interviews

工作室十周年之际访谈王振军

青锋老师： 10年，对一个团队而言的确是个重要的节点。我学习了关于您本人和工作室之前的资料，读了您的《蔓·设计》，我觉得您的有些想法、观点还是挺有意思的，所以很想和您讨论一下。因为我在清华大学教建筑历史和理论，所以对理论问题也很感兴趣。看您在《蔓·设计》里边讨论了很多这方面问题，我就想把她结合到具体项目来讨论一下。

王振军： 工作室是2009年成立的。到今天整整10周年了，好多朋友都说我们无论如何应该纪念一下。

青锋老师： 首先有一个问题，就是说在一个大型央企设计院这样偏生产的环境下，您是怎样做创作的？因为我们讲的比如像路易斯·康和赖特，他们都是在事务所模式下的。我看您这儿原创作品占比很大，您是怎么做到的？

王振军： 这桌上的两个文本都是我们今年在做的项目。一个是紫光南京集成电路项目，另一个是新疆阿克苏三馆一中心。项目类别、规模和收益都完全不同。但我们一视同仁，都把它当作创作的机会紧紧抓住。

清华紫光投资的这个存储器芯片项目，投资750多个亿，几乎与大兴机场投资相近。这种超高造价的项目在近些年的高科技工业项目中很常见。当然工作室指标完成主要是靠这些大项目来支撑的。说到紫光这个项目，甲方副总对设计的价值极为看重，他说："王总可别当一般的工厂来做，需要您能做出'奔驰'的感觉出来。"所以即使这是一个生产性项目，在甲方的支持下，我们一开始就把它当一个精品来做。我们还做了多个不同比例、材质的实体模型反复推敲。

相比来说，阿克苏项目规模小很多、设计费不高。但我们同样全力以赴。我最近在提一个观点就叫"大建筑观下的蔓·设计"，就是说要淡化民用建筑和工业建筑的界限。其实这是针对我们这样一个工业基因很强的设计院提出的，身处其中的建筑师在生存同时如何去保持创作初心。回顾现代建筑史的发展历程，现代四个大师中的三个都是拜贝伦斯为师，从工业项目例如著名的透平机车间（现代主义建筑的第一个作品）开始职业生涯的。

我一直主张模糊创作和生产的界限。这30多年也确实要求自己不论工业建筑和民用建筑都认真去做，不能说都得奖，但至少不要有太多遗憾。从2009年做工作室开始应该算一个节点，主观上已经完全不存在生产和创作之间的界限了。

青锋老师： 因为我也接触过一些大院的总建筑师，他们基本可分成生产型和创作型两类。您可能是处于一个更理想的状况。我发现您这里有很多特殊的、专业性很强的项目，如机场塔台，它会有很多特定的限制，对创作约束很多。但我看您做了很多，还很有特色，以这种类型来说明，像这种限制这么多的项目，您有哪些障碍需要去克服？

王振军： 机场塔台在空港项目里属于那种规模不大、要求很高但收费很有限的项目，相比大家更愿意做航站楼、机场跑道等高利润的项目。

因为首都机场T1塔台，是我院前辈1977年做的。到1997年T2塔台时，我觉得应该把前辈们的业绩传承下去，最后通过竞赛就拿到了合同。说到塔台设计的难度，首先它是机场这十几平方公里范围内最高的标志性建筑，因此引人瞩目。再者顶部的型制是国际民航组织规定的，全世界都一样。比如这个明室玻璃，为了防止反光一定要向下收18°。从贝聿铭在20世纪60年代开始做的美国塔台系列就是这样。然后就是它整体剖面的限制，上下大中间细。还有工艺及公共设备系统的复杂性。在限制多的同时，城市部门和机场当局对他的艺术性要求又很高，像我们做T2塔台、T3塔台，都要在首都艺术委员会过会谈论的，审查规格很高。T2塔台我们结合顶部的形制融入天坛意向，想让旅客一抬头看到它就想到这是到了北京。结构抗震上其重要性等级等同于人民大会堂，就是北京如果发生8级的地震塔台还要能使用，所以结构设计也费了很多精力。还有智能化系统、公用设备系统和消防系统在国内当时都是最先进的。做这个塔台很能锻炼一个队伍的综合能力，它不光要求建筑师厉害，结构、设备，包括智能化系统都要紧跟国际的先进技术和标准，这种项目其实正好吻合了我们中国电子院的一个特点，我们每个专业可能拿出去不是最强，但是综合起来我们是最强的。

之后我们又中了T3塔台的标。本来福斯特想做它，但因为这是一个保密项目没做成。能够中标就是因为我们的设计概念——塔台和T3航站楼互为投影，把T3航站楼平面给立起来就是塔台，反之亦然。另外同时它像中国古代的酒器觚，寓意了机场这一迎来送往的庆祝场所。福斯特设计事务所非常赞赏这个构思，建成后获了很多奖。之后我们做了天津塔台，天津塔台当时的构思来源于天津市花。还有西安塔台我们是想让它有古代密檐塔的那种感觉。郑州塔不同于以前的塔台，它不在机场轴线上，而是偏在一侧，这就提供了做出不对称塔台的可能。后来我们到河南省博物馆采风，讲解员介绍到了镇馆之宝——贾湖骨笛，说是7000年前人类最早的乐器。由此引发了我们的灵感，其形状和功能与塔台很匹配，同时又和当地文脉取得了联系，所以，后续方案汇报审批非常顺利。另外，这个塔外墙为清水混凝土，现在它号称是中国最高的非线性清水混凝土建筑。

青锋老师：我觉得在您这几个塔台的设计里，可以看到很典型的一种设计策略，就是对"建筑意"的内涵把握和当地文化元素的采集。您做的沙特馆，其实也是类似的做法——来自于"月亮船"的概念。我当年想去参观，但因排队太长只好放弃了。我觉得沙特馆是一个非常典型的博览会建筑，它要在很短时间内很直接地向观众传达信息，用大众比较能够理解的一种符号或者一种场景把它传递出来。这让我想起了之前我看到文献里面的文丘里1992年给塞维利亚世博会做的美国馆，极端到什么程度，它就放了一整面美国国旗来做立面。所以，不管沙特馆还是这几个塔台，我觉得可能最成功的还是您找到了这个想法（idea），为沙特甲方和大众所接受，这点上您有什么诀窍？怎么获得大家都比较认同的想法（idea）？

王振军：做沙特馆是在2007年的11月份，当时一个朋友说在网上沙特城乡事务部在通过国际竞赛征集上海世博会沙特馆的方案，没有报名门槛。当时我们正在设计长安街上国际财源中心（IFC）。到了晚上我研究了一下标书，第一轮对文本要求不少于5张A4纸就行。报完名我们就利用业余时间，做了很多方案，最后我们出了三个方案，有两个进入第二轮，"月亮船"最终胜出。

从一开始的放松投标、中标，到建成、到最后等待开幕的过程，特别是到后期其实是一个逐渐煎熬的过程。项目开工以后，由于沙特方的休假特别多，本来比较充裕的施工周期，到最后却非常紧张，一直到4月30号，我本人因为血压高的都差点出问题。再一个压力就是在这样一个国际性的建筑秀场上，我们将会得到怎样的评价。那时候英国馆、德国馆、丹麦馆陆陆续续公布方案并落成，业内外媒体对欧美馆热情高涨、大肆报道，这时候沙特甲方就总在问我"王先生，我们能排到第几？"世博会开幕后的第二天，沙特馆门口出现排队现象。因为世博会参观通道栏杆数量有限，会务方就把栏杆都摆在预测热点欧美馆。第三天会务方才把参观通道栏杆搬过来。那张照片是5月2号傍晚用卡片机拍摄的，5月3号摆放上排队栏杆后，因为栏杆上带帐篷，再想拍高质量的照片就找不到角度了。

这个过程我觉得对一个建筑师成长来说，特别历练、非常难得、很有意义。一开始你自己也会疑惑：建筑或者说世博会建筑到底应该是什么样的？观众到底会喜欢什么？站在T台边等待检阅和比较，这个过程的确是非常有趣也非常忐忑的。

由于甲方效率问题导致最后效果有很多遗憾，就说一点，我们原来这个船的外表皮是镂空的伊斯兰经典纹理金属板，里边用水泥砂浆抹面，然后再打上灯光，效果非常好，但由于加工时间不足，最后只能用普通的蜂窝铝板把它包裹起来。

通过这个项目我总结有几点：首先业界对一个建筑的评价当然很重要，但大众的喜欢或者说"票房"才更重要。而且两者之间也是有共识的。那么什么是好建筑呢？我觉得第一个，要感知度比较高。就是刚才您说的idea，这个概念表达你要让专业的建筑师能看懂、老百姓也能接受。一个建筑专业评委说特别好，老百姓却特别不喜欢，这种现象好像很少。说到沙特馆的感知度，当时标书上有两句话，一是上海世博会的主题——城市让生活更美好，二是要反映中沙友好，我们自然而然就想到从中沙友谊引发的海上丝绸之路，其载体就是船，这一点毫不晦涩、通俗易懂。另外我们的总图将船头转了个角度朝向麦加，让沙特的部长特别感动。

第二个空间是主角，做建筑最重要的还是在做空间体验。我们为什么把这船架起来？在设计等候空间的时候，我们就查了以前的资料，说世博会场馆排队最长记录是3.5小时。所以我们就考虑观众从大门进入展馆，可以慢慢走，在排队过程中可以观看架空舞台的表演，船体架空也为观众营造了一个相对凉爽的小气候。通过坡道进入展厅以后，我们利用这个船的内壳做了一个屏幕，在吊顶上放置了 26 个投影仪，同时投到这个壳形的屏幕上，我们当时受《一千零一夜》里阿拉伯魔毯故事启发，用一个弧形的自动步道架在屏幕上，营造出来一种全新的、沉浸式的观展体验，取名叫"全景立体融入式参观方式"，利用全新的空间体验打动了观众。这就体现了所说的空间体验要生动感人。

第三个就是形式和内容统一。设计了这个船，然后利用这个船去设计新的参观流线，如果说你设计了这个船，但船里边还是利用楼板去做分隔，那这个建筑就完蛋了。

第四个就是关照文化的传承。沙特馆用月亮船、阿拉伯魔毯、伊斯兰花园加贝都因帐篷等沙特人文化符号获得了甲方的认可和观众的喜欢，就很好地传播了沙特文化。

青锋老师："感知度"就在于它一方面确实要有比较突出的形态，另一方面它能够被理解，这是建筑适合向大众传播的一种特质，在沙特馆这个项目上可以说展现的淋漓尽致。另一方面像您说的内容和形式的密切结合，我觉得您在谈到"蔓·设计"的观念时是另外一个重点，因为我理解"蔓·设计"里的有机建构，它最核心的一个理念就是它强调的是由内而外和由外向内的过程。更早之前路易斯·沙里文的"形式追随功能"其实也是这个观点，就是在建筑的形态、材料、细节、空间等方面，都是要去结合它的功能。提到有机建筑理论，这是很有趣的一个话题，因为如后现代、粗野主义等有不同的理论倾向，您是在什么样的状况下对有机建筑理论投入更多关注的？

王振军：我觉得提出"蔓·设计"有一个最触发我重新思考的点是，大概在二零零几年，在北京，有一次我开车从机场回来走北三环，目之所及的建筑一个比一个让人沮丧。按说北京作为首善之区，建筑师占比在全国应该是最高的，而在北三环这样的城市主要空间带上，怎么会有这些建筑？后来我在一本书里看到一句话，大意说人类再伟大，你建造的东西现在看来还是不如天然的那么和谐、优美，包括我们现在所倡导的绿色建筑、可持续发展，其实都是说明我们以前做得不好，现实需要建筑师去反思。我觉得，造成这样的结果不是建筑师的能力、技巧问题，而是观念出了问题。

工作室从成立到现在已经 10 年了，不仅从现实需求考虑，还因为院里给我一个很重要的任务就是培养年轻建筑师。怎么去培养他们、怎么让他们走一条比较健康的创作之路。你要把自己的理念告诉他们，又不可能长篇大论地说，需要把它总结成比较精炼的内容。于是，我把这些年的思考，包括沙特馆等项目的经历及感受总结出来，把关于"有机建构"理论、建构文化研究等内容消化、凝练下来。另外，结合建构的概念把赖特的理论进一步拓展，从方法论中总结、打造高完成度建筑的办法。其实，建构概念在工业建筑中早已使用，它对各个专业的配合精确度要求是高过民用建筑的，只是自身的意识在这些方面还没有觉醒。在此，我重新梳理一下：首先，现实的城市景观告诉我们，只有建筑师强大没有用，你要有一定的觉悟，要有一定的认知高度来重新审视建筑这件事。第二个就是建构的问题，比如说你看一个植物的组合关系，就是一种建构，包括我们工业建筑，在整个生产、投入、使用的过程中，从各个专业之间密切的配合关系出发，我在想是不是应该总结出一套可以告诉年轻人的在日常的创作中应该遵循的原则。

那么为什么用"蔓"字呢？包含几个意思，第一层意思是建筑要像自然界那样有机建构，有机的东西才是具有生命力的。另外一层意思是要把这种哲学思考，蔓延到创作的全过程中，把它做成一个闭环的东西，现在几乎所有项目都是设计总包，现实已经在走这一步了。这只是一种哲学思考，还在补充、发展和总结过程中。但至少现在我们工作室这些同事们，说起"蔓·设计"大家都心领神会。

青锋老师：它是一种设计态度，需要内生性去贯穿始终的。当然这里提到赖特，我觉得有一个很有趣的差异，您这里面也提到，赖特设计的建筑尤其在城市里是非常封闭的。他早先的不管是拉金大厦、唯一教堂，甚至到后来的古根海姆博物馆，都是很封闭的，因为他确实不喜欢城市，所以他一定要把建筑和城市隔绝。但是您的很多项目都是在城市环境里，却都呈现出一种更开放的态度。信达的合肥灾备中心、海盐的创新中心，包括郑州空港小学，都明显得可以看到是从分析一些城市因素开始的这样一个很清晰的流程，通过这个流程来展现您的理念，从产生到贯彻，到最后的结果，这个流程我觉得实际上体现的是一个设计体系，从大的、抽象的到一个相对比较具体的，甚至最后到细节和室内的过程。

王振军：我为什么喜欢讲沙特馆，并不是说我多么在意这个项目，而是这个过程给我们带了很多启示，不管你是中国的客户也好，外国的客户也好，其实这套方法如果得当的话，你做的一定会非常成功。所以现在我们做项目，包括投标、讨论方案，其实都是按"蔓·设计"这套方法实施。刚才我说的价值判断，什么是好建筑，为什么要选这个方向而不走那个方向，其实都是用这些原则来判断的。

青锋老师：对那种有一套确定的、相对比较固执的建筑语汇的建筑师，您怎么看？

王振军：不是说建筑师有两种嘛，一种是刺猬型的，一种是狐狸型的。这个比喻比较形象，扎哈肯定是刺猬型，"蔓·设计"的主张应该更倾向狐狸型吧。这种分法包含了很多含义在里面。建筑学也像其他专业一样都是有他的自治性。而不是说现在是信息时代了，是所谓的"泛中心化"，没有权威了。现在看来，我认为对建筑学还是有共识的。我们最近和很多专家一起评一些建筑设计奖，好的建筑最后得票还是很集中的，这就是共识。在共识之下每个人可以再讲究自己的特点，建筑师也需要这么做。最近我的一个学生到日本读博士，据她说日本高校考核老师要求每年要出一本书。只要有自己的观点就行，无需特别严谨。所以你看日本的建筑界，比如在普利兹克奖上，日本会不断地出现新人。他们建筑创作的共识是放在那儿的，只不过每个人的切入点不一样。从丹下健三到安藤忠雄、妹岛和世，他们还是在搞空间、材料、光线这些东西，只不过切入点不同而已。我觉得之所以想把"蔓·设计"总结出来，就是想把这些有共识的"评价标准"放在那儿，然后每个建筑师把她融入到自己的路径或者切入点中，一定会做出有特点的东西！

青锋老师：您说的我很认同，共识也是我们老师教建筑课程的一个基点，共识的东西一定是可以教的，而每人自己个性的一套就没法教。我觉得在您最近几年的作品中，有一个概念还是很突出的，这也是比较有共识化的概念，您称为"自然场力"。我们刚才讨论的城市乱象，就是因为建筑师太不在乎城市文脉造成的。而像您设计的不管是很大的园区，比如海盐创新中心偏重景观，还是合肥信达向城市打开一些开口，将路径引进去，而郑州空港小学则是思考和城市包括和整个学校的构成关系的体现，确实像您说的是一个"狐狸型"的设计策略，很明显您对自己还是有明显的约束。

建筑师要利用这些自然要素去影响它，这也是赖特有机建筑理论的重要观点。有句话说建筑师是一个编织者，而不应是造物者。是编织者，就是去把已经有的东西利用、组合起来，让每个元素都处于最好的状态，自然而然地形成一个和谐的整体。在你们最近一些项目中，也可以看到这个特点，从每个单体或每个部分可以看到它的个性还是很强的，但是因为有场地因素的引入，又觉得总体是和谐的。上海浦东软件园三期您将场力因素引入，使每个单体的特色都很鲜明，组织在一起又显得很整体。

王振军：是的，比如说郑州空港小学，因为它是个集群设计，请了全国十几个建筑师，我被分到一块地设计小学。设计建筑我比较喜欢从场地、空间、场景开始构思。我想起自己的小学时光，那时候没什么可玩的，男生课余就玩斗拐。男孩精力旺盛，加上学校也不大，从东到西，从西到东，疯狂地玩。所以，我希望这所小学应该搞一个可以供学生活动的连续空间，把各个功能区连接起来，而且它是通过坡道立体化地连在一起的。我就想着孩子们下了课可以斗拐从屋顶上斗到广场。甲方很喜欢这个方案，他们也觉得现在小学生作业任务太重，导致课余时间被严重挤压，很需要改变。

青老师刚才说的"编织"这个词来讨论科技园项目我觉得很贴切。大概是 1993 年 ~ 1994 年，我们开始设计中国最早的软件园——浦东软件园。那时候没有资料和规范，甲方总经理从美国带来很多硅谷的信息。但是硅谷是一个城市区域自然生成的概念，而科技园区是非常有中国特色——政府主导的一个项目。说到编织，软件园从一开始就有这一典型特征。建筑师的责任是把土地、建筑和景观、人才、技术、资金引入，然后编织在一起，变成一个高品质的物理空间。浦东软件园一期，甲方不敢投太多钱，也没有钱，之所以中标，是因为我们的被动式绿建概念。项目有 24m 限高，容积率是 1.4，这样密度非常高。我们就把餐厅的屋顶和园区的集中绿化带利用坡度整合到一起，人员从主入口进来，6% 的坡度与餐厅非常自然地连起来，感觉园区的绿化率被放大了许多，是个很通俗的想法。

浦东软件园三期 50 多公顷，位于上海张江的几何中心。一、二期的成功使甲方赚到了钱，所以甲方想要一种国际化风格。这块地的上位规划被十字路一分成四，我们的规划结构是在与城市接口不变的情况下做了个环路，一来可以降低车速，二来减少过境交通，第三是通过环路把这个园区分成了三个层次，一个是外侧靠城市主路的是大企业，然后靠环路的是中型企业，在中心湖和环路内侧布置小企业和一些公共空间，这样的规划结构获得了评委会一致认可。

海盐中欧合作园，它的特点是第一离海特别近，第二个它在研发下面有一些中试车间。我们总的构思是要和海从视觉上、路径上取得联系，让员工很容易地看到并到达。把一些建筑局部架空，让出一些地面空间给人们活动，所以我们就将园区分成道路系统、景观系统、建筑系统、还有水景系统，最后把它们叠加到一起。

正像刚才您说的，创新一直是我们的追求。在设计这个园区总部大楼时，我们提出设计一个不一样的主楼——它有轴线、是平衡的，但又感觉它是变化的。因为这个园区位于中欧合作大片区里，我们用合作（Corporation）和信心（Confidence），两个 C 组合到一起，这样构成感很强的主楼就被设计出来了。非常规矩的两个矩形组合成 L 形海景办公空间，将杭州湾一览无余。这个园区我们还有一个概念，叫"风从海上来"，建筑在靠近廊道两侧做些变化，感觉像有风吹过一样。而靠近城市一侧，就让它理性厚重一点。而园区节点上的餐厅等，则都做成非线性的，既放松又浪漫。

青锋老师： 沙特馆因为它是一个博览会建筑，所以在感知度方面可以设计得夸张一些，但是其他这几个项目，我觉得从建筑学角度来说您的控制手段更成熟。比如两个 C 组合，这些都是比较容易理解，另一种就是建筑和景观的结合、城市脉络的引入，这都是经典的建筑学对于场地因素的引入策略。这也是当年所说的双重译码，它指的是符号或密码，你可以理解为一种是经典的建筑学传统，一种是大众化的。这点其实文丘里的建筑体现最鲜明，他有一些经典的拱、山墙，但是它也是大众符号。经典的和大众化的都可以扩得更广，不一定只是落在符号上，也可以落在各种层次这样一种情景或更深层次的脉络上。

我觉得像您的几个这种大型的园区项目，在规划结构、空间体量、材料选择、比例尺度等方面都显得越来越成熟，我们都可以从中看到一些经典的或者创新的构思，这些东西合起来，我觉得确实比那些传统的园区、建筑，更有内容和活力。

您在大兴机场设计的几个项目我也看了，空港的文脉条件是挺枯燥的，在一个个方块街区里，但是通过元素编织或者编制故事，您们还是能用各种大众化或者经典的元素把它建构起来，也同样表现出很充沛的内容，而且还有很多细节。我觉得这对于尺度很大的建筑或者建筑群，您们的实践是很有启示意义的。

王振军： 在投标北京大兴国际机场时甲方都会给你一本该机场的城市设计导则，包括贴线率、高度控制等要求，它的主要目的是为了突出航站楼，让其他建筑都成为背景。原则上没有错，但是这种导则如果太定量化，会给建筑创作带来很多束缚，很容易使建筑变得平庸。

我们的这些建筑最后获得好评的原因，主要还是因为在严格的限制下有所创新。比如生活服务中心，为呼应高架桥返回北京的车辆，我们就把这个合院的东南角改成了圆角，然后在院子入口做了处理，强调了主入口。这就是"蔓·设计"所倡导的发现自然场地、挖掘深层结构所得到的结果。这都是在中标以后我们做的进一步的优化。这个园区住的都是年轻人，建筑一定要有点活力。那么我们就通过这些绿色建筑的做法，从剖面、灰空间、空间和形式上尽量做得生动丰富一些。我们还有一个得意的理念是"间间有阳光"，这也是我们中标的最主要原因，整个宿舍区没有北向房间，东西向房间通过小阳台和深窗套来解决东西晒问题，专家一致认可了这个理念。可以看出来，我们的设计始终都在讲逻辑，但不是说讲逻辑就没有艺术性。逻辑之下就要靠想象力和创新的冲动，所以我们在招学生时会特别关注他们的艺术感觉，这对建筑师来说是非常关键的。

工作室十周年之际访谈王振军

青锋老师： 说到招学生，我们建筑学院老师也很关注，需要培养什么样的建筑人才，或者说我们建筑教育要去培养哪方面的素质？最有发言权的应该是像您这样的雇佣者，这方面您有什么建议给我们？

王振军： 我们应该叫接收者，关于建筑教育大家在一块议论得还是比较多的。我一些同学的孩子是学建筑的，有次聚会我就问他们喜欢哪个建筑大师？皆答：扎哈，且没有原因。我和他们的父母都认为这样教育结果会出问题。过于关注形式容易导致对建筑学理解的偏颇。

1986年本科毕业的时候，我为什么后来选择留校，就是觉得建筑学涉及的内容太多，四年本科并没有把建筑学理解透彻。比如中国古建史和国外建筑史也只是学了个大概，感觉好多内容刚刚开始学就结束了。其实这些课对建筑师设计观的形成影响是特别大的，所以我就想在学校里再待几年，当助教之余，花时间再把那些中外大师的作品认真研究一下。两年后虽然读了研，但直到现在从事设计职业30多年，都还觉得建筑学还有很多问题没想透，建筑学除了本体的，还有外延要素，另外还有表达手法的问题，建筑学要解决的东西太多太多。所以我希望现在的学生能够去关注建筑学深层次的东西，而不是仅停留在表现的层面。现在回想起在教室、宿舍里讨论这些国际大师，大家各抒己见，通过讨论和争论，对理解教科书里的内容非常有意义。现在我能感觉到年轻人的聚焦度在下降，这可能和信息时代资讯的泛滥有关，老师这时候更要多加引导，不能仅停留在浏览层面，而应去深度解读。我觉得本科的时候对我影响特别大的就是李峥嵘老师，他讲近现代大师的作品课的时候把那些作品描到胶片上，然后用投影仪播放，给我们讲这些建筑空间的体验，空间的开始、铺垫、高潮等，使我受益终生。

最近这些年我感觉招聘进入工作室的学生平面功底比较弱，但这是基本功，如果没有这些基础训练，你说你想设计出好作品，那是不可能的事。现在看路易斯·康、贝聿铭那些作品，我们还是没有超越。所以我觉得建筑教育把学生的基础打好非常重要。

青锋老师： 您说的我特别认同，因为我现在在清华大学教近现代史建筑，包括我们正在编写近现代史建筑的新版教材。我刚才聊到赖特、路易斯·康、文丘里，虽然快100年过去了，但建筑学的基本逻辑还在，它构成了整个建筑的一个核心的价值基础。另一点建筑学作为一级学科，是有自己自主性的东西的，我也比较强调要深度去理解，对大师，不仅是要对他的建筑作品进行研究，还要关注他们的理念。虽然现在建筑变化很快，但那些经典思想还是一直存在的，这一点也体现在今天我们交流的很多作品上。您的作品也都有一种轻松感在里面，一方面容易理解，另一方面可以看到很多经典的和大众化的手法。不管是把"蔓·设计"叫作理论还是叫作设计理念，当一个团队有了自己认同的思路或者原则，做设计时就不会盲目；另一方面有相应的操作方法，您也会避免纠结。有了架构在那儿，您就会深入到方法论的探索，也会找到更多不同的切入点。

您的这些项目、包括建筑教育观点，我觉得有很强的系统性。就像开始时说的"蔓"的概念，它是由内生发的、有机的，但关键还是她的整体性，从策划、设计开始，到最后建构结束，要把各个因素整合起来。我觉得"蔓"还是很鲜明地体现了您10年来作品的共同特征，这对其他建筑师是很有启发意义的。尤其是在大型建筑实践方面，比如像园区或者空港建筑，不是说把一个小建筑简单放大。面对大尺度和复杂内部组织，去引入这样一个理论来作为思考框架，是很有启示作用的。最后我想说：很期待看到您更多、更新的作品，去体现和验证"蔓·设计"这样一个很好的设计理念的价值所在。

青锋：清华大学建筑学院副教授，博士生导师，建筑系副系主任，《世界建筑》杂志副主编。主要研究领域包括现代建筑史、当代建筑设计理论、建筑与哲学、建筑教育以及中国当代建筑评论。其以历史与理论为背景的学院式评论工作获得建筑师与研究者的广泛肯定。在国内外重要学术期刊发表论文数十篇，参与编写了《外国近现代建筑史》等教材，出版有《在托斯卡纳的阴影中》以及《评论与被评论：关于中国当代建筑的讨论》《当代建筑理论》等著作。目前任中国建筑学会建筑传媒学术委员会副秘书长，建筑评论学术委员会理事。

我的建筑认知体系：蔓·设计

提要：

他于湖南大学毕业后留校任教两年，随后读了研究生，在学校九年的沉浸对他后来的建筑设计生涯有很大的影响。1991年他被分配到中国电子院工作，从基础画图一步步走到今天总建筑师的岗位。2009年王振军工作室成立，近来他提出"蔓·设计"的创作理念，这也是王振军30多年建筑创作的思想总结和基本认知框架。他认为对"什么是好建筑"业界已有共识，接下来在这一共识下建立自己的认知体系是未来建筑创新的潜力所在。

本期嘉宾：王振军

中国电子工程院设计院集团股份有限公司总建筑师、建筑专业委员会主任，王振军工作室主持人、研究员级高级建筑师、国家一级注册建筑师，中国建筑学会常务理事，北京建筑大学硕士生导师、国务院特殊津贴专家、2019央企劳动模范

张力： 请您介绍一下求学和进入中国电子院之后的大致经历？

王振军： 我是1986年湖南大学本科毕业，毕业之后其实很想去深圳，因为当时深圳刚开发，有很多发展的机会。当时建筑学还是四年制，我又觉得建筑学涉及的学科太多，毕业时有很多东西还没有学透，加上在老师的挽留下，最终留校当了助教。两年后又读了近三年研究生，所以我在学校前后待了九年时间。虽然当时建筑学资料比较匮乏，但是由于学习时间比较充分、目标聚焦，所以通过九年在学校的沉浸，我可以从深度和广度上琢磨涉及建筑学的繁杂的知识信息，所以这九年对我现在的职业影响挺大的。1991年我被分配到中国电子院，一直工作到现在。我们很多同学都跟我开玩笑说：你再不走都变成"文物"了。或许我的性格就是这样，不太喜欢折腾。很多同事、同学离开设计院之后都不在设计一线了，但是我本人还是比较喜欢设计、画图，所以就一直留在了中国电子院，而且还在创作一线工作着。

张力： 您到中国电子院后，建筑设计工作是从哪个阶段开始的？因为您建筑读了九年，是从做方案开始？还是从画基础的图纸开始？

王振军： 我被分配到中国电子院跟其他的毕业生是一样的，都是从基础画图开始的。刚到中国电子院的时候，我一开始是有点困惑，因为中国电子院在二十世纪九十年代是很传统的工业院的体制——工艺主导。当时不管做工业建筑还是做一个跟工业建筑相关的民用建筑，建筑师创作的自由度都不是很大。但通过这么多年的发展，我觉得中国电子院建筑设计现在已经发展得很好。中国电子院在一些发展比较早的特区城市比如：深圳、海南（最早成立分院）、大连、上海都有自己的分院，这些分院主要是做民用建筑。建筑骨干经常被派到分院工作，我在深圳分院工作了两年，在海南分院工作了三年。首先是做设计，接着是做专业负责人、项目总设计师，后来技术职务是主任建筑师，最后是总建筑师岗位，一干就是快三十年，我这三十多年就是这么一步步走过来的。

张力： 在这些分院的工作会不会比总院更加接地气？遇到的困难也更多一些？

王振军： 分院的工作非常接地气，经常会有一些短平快、突击的项目投标，我现在认识很多圈里的老总都是当时在一起竞标过项目的，我觉得跟这些高手交流收获挺大，在分院的锻炼价值特别大。我印象比较深的是在海南分院工作期间设计的五星级山海天大酒店，是鲁能投资的项目，1996年开始设计、建设，1998年建成使用。当时我们在黄星元总、张建元总的带领下与众多国内外建筑师同台竞技，拿到这个项目的设计权，从建筑设计到室内设计全部由我们这个团队来完成。

张力： 当时为什么会成立这个工作室？

王振军： 我的工作室是2009年9月份成立的，到今年正好10周年。当时我担任院里的总建筑师，院领导对建筑设计非常重视，对工作室也提出了具体要求：第一，做好院里总建筑师的技术管理和重要项目、科研的评审工作。第二，做建筑精品。不要求出数量，但要出精品，给中国电子院的建筑师做一个引领，使院保持一个比较高的建筑创作水准上。第三，培养年轻建筑师。吸收刚毕业的研究生和年轻的建筑师来加入到我们这个团队，在这里会有长期或短期的培训。通过这10年的培育，有一些优秀的建筑师留了下来，而且现在已经成为中国电子院的建筑师骨干，这点是我最高兴的。

张力： 进入您的工作室有什么要求？

王振军： 从某种意义上讲进入工作室还是有门槛的。首先要酷爱建筑学，这样才能经得起外部诱惑，才能坚持做好建筑这个行业。因为建筑师的培养是一个很漫长的过程，如果做了一段时间就坚持不下去了，意味着培养过程的中断，这样也很影响成材率。另外就是强调要有团队意识，因为我们要高效率地做出建筑精品，这是要靠大家一起通力协作才能做到的。

张力： 工作室的培养体系是怎样的？

王振军： 我觉得还是师徒制。比如我是工作室主持人，我下面有四、五个设计总监，他们都有自己的一个小组，他们直接带年轻的建筑师，我们之间形成一种师徒关系。遇到重大的技术问题、重大的项目我会提概念和把方向。

张力： 他们的方案您也会参与创作吗？

王振军： 我们方案创作是非常开放的。我自己做方案，同时各设计总监团队也做方案，工作室投票，不论是谁的方案大家都可以畅所欲言来质疑，当然我的方案也有选不上的时候。未来我想往后再退一退，有些项目我可能就不用再自己去做方案了，只参加讨论就可以。我除了自己保持一定的创作量以外，主要是给工作室年轻的建筑师创造更多机会，我认为这个人才梯队建设对设计院是最关键的。

张力： 您在办公建筑、科技园区领域都有精彩作品，请您给我们介绍一下比较经典的项目！

王振军： 我印象最深的第一个项目是新世界中心。这是邓小平同志南巡回来后北京第一批招商的项目，投资商是香港新世界的郑裕彤，由我们和香港伍振民事务所（现在叫刘荣广伍振民工作事务所）合作设计的。新世界中心有 20 万 m²，是我从 1991 年进入中国电子以后接触的最大的一个项目，而且是一个特别复杂的城市综合体，里面有酒店、写字楼、公寓、商场，还有一个特别复杂的人防系统，而且项目离天坛不到一公里，处在古建保护控制范围内。项目设计之初我们到香港参观了很多优秀的综合体建筑和知名的事务所，让我们对香港一些比较先进的职业建筑师的工作方法、理念、创作手段有了近距离的了解和学习。当时整个北京城市建设还比较滞后，但是与香港建筑师合作设计的这批建筑在建成后，一下就把北京市中心城区的建筑品质提高了一个档次。

我很有幸参加了新世界中心的全部设计与建造过程，后来这个项目被评为"北京九十年代十大建筑"，还被作为古城建设升级的范例。我觉得新世界中心项目无论从专业追求上，还是它建成使用后对城市的贡献上，都做得非常好。通过新世界中心项目的实践让我对大型城市综合体这类建筑的设计积累了宝贵经验。

以这个项目为契机，后来我也做了很多别的大型项目。比如长安街上的财源中心（北京 IFC），项目地址在长安街上，靠近国贸，是我们院和加拿大宝佳国际建筑师有限公司合作完成的。财源中心（IFC）是一个超高层建筑，150m 高，由 4 栋超 5A 智能商务写字楼组成。项目地块是 280 多米长的扁长形。因为长安街地段比较敏感，我们就先做城市设计，通过分析整个长安街上各种高层建筑和它与左邻右舍的关系，并从城市的韵律、功能布置、日照关系上进行分析，包括对建筑风格的研究，都做了大量的工作。当时我们还专门到德国去参观他们的金融区和经典的金融中心写字楼。

我这里还有一块业务是机场塔台建筑。由于机场塔台比较难做、规模不大，但导航工艺上又有很多限制，由于它是空港区里的最高建筑，所以规划部门对它的艺术性要求比较高，尤其对形象特别关注，很多设计院觉得麻烦不愿意做。我当时有一股执着劲儿，别的设计院不做，那我就去做。正好有这个切入点，后来我们投标中了首都机场 T2 塔台。接着又做了 T3 塔台。

机场塔台是一个城市门户的标志，方案评审都要上北京首艺委的会。给我印象比较深的是 T3 塔台的设计。T3 航站楼长度是三公里，建筑面积将近 100 万 m²。T3 塔台，建筑面积不到 3000m²（2885.38m²），高度是 82m，这是完全悬殊的两个体量，但是，规划局要求塔台和航站楼要有呼应。后来我们通过不断的讨论、思考，就提出了一个概念：把 T3 航站楼立起来，再缩小，就变成塔台的雏形，以此作为一个母体来进行创作。由此塔台的造型和航站楼非常巧妙地取得了呼应。我给它起了个概念叫：相互呼应、互为投影。

T3 塔台审查时跟航站楼的后期如景观、建筑照明一起上会的，福斯特也在场。我们当时还提出了一个概念：就是 T3 塔台的形状和青铜器里面有一种酒具在商代叫"觚"的形状是一样的，而塔台是作为一个城市的门户，正好跟中国的酒文化非常吻合，迎来送往、悲欢离合。加上"互为投影"的概念，一提出来，规划局、首艺委、福斯特都很满意。

塔台虽然面积不大，但是因为它的工艺比较复杂，特别是明室导航台要求360°无障碍的观察视角。所以我们在建筑、结构以及设备系统上想了很多办法。这个项目最后获得国家银奖。这很不容易，因为塔台的建筑面积小、体量小，别的项目都是几十亿、几百亿甚至上千亿投资。这个项目告诉我项目不分大小，只要尽心去做都会有好结果。

T3塔台做完之后，我们又做了上海虹桥、天津、西安、郑州等一系列机场塔台。我现在和民航局空管处一起编制塔台的规范，可能明年就发布了。

张力： 请您跟我们普及一下塔台的设计对机场的重要性？

王振军： 塔台是机场重要的运行指挥中心，是专业性很强的建筑。飞机在起飞前的滑行、起飞，包括回来飞机的降落、滑行到登机桥，主要由塔台来指挥。所以塔台对视线的要求、智能化的要求、结构抗震的要求，都非常高。比如首都机场塔台就是按国家生命线工程来做的，比一般的重要建筑还要提高一个等级。如果出现地震等灾害，塔台是不能出问题的。

张力： 请您讲讲您创作的2010上海世博会沙特馆吧？

王振军： 沙特馆项目是世博会上外国馆里唯一对建筑师完全开放的一个国际竞赛项目。当时任何人都可以参加，没有门槛要求。偶然的一个机会我们参加了沙特馆的竞赛，很幸运的就被选上了。当时沙特这个国家在世博会上不是很受瞩目，所以它由不受瞩目到被关注，再到最后变成一个特别受追捧的场馆，其实也是我对建筑创作一个思考的过程：就是到底什么是好建筑？建筑师创作应该要追求什么？我印象最深的就是，当时42个国家的馆同时在建设，有的馆已经建出来了，比如英国馆，外形是非常酷炫、非常神秘的那种造型。包括其他国家的一些馆的造型，也十分引人瞩目。我当时心里很忐忑，压力也很大，而且甲方也给很多压力。沙特馆建完后会不会受大家欢迎？后来我在一篇文章里写道：其实世博会就是一个建筑秀场，所有的馆在同一个T台上比较。世博会开幕以后，我们坚持的创作原则，都得到了大众的认可，我觉得这个是很重要的。因为有些馆可能在建筑界、学界评价特别高，但是参观的人特别少，一位前辈曾说，最好的建筑就是要得到观众的认可。做完沙特馆之后我体会到这句话是对的。我们是设计总包，监理是英国的RSG，展陈设计是西班牙GPD，施工是中江国际，加上甲方沙特城乡事务部；这五个团队工作将近三年，成本是挺高的，如果只做了一个建筑形象是没有意义的。后来很多同行去参观了沙特馆，对团队坚持建筑学的本质创作，并同时被大众所喜爱所产生的效应也给出了很高的评价。

张力： 沙特馆当时为什么这么火？您认为最关键是哪几点？

王振军： 关键是创新！在42个外国馆里，它是唯一从"参观方式"上进行原始创新的馆，我们把它叫"全景立体融入式参观方式"颠覆传统的参观方式。当时沙特馆的竞赛标书有两句话要求，一个是：城市让生活更美好。这也是整个世博会主题。另外一个：要体现出中沙友谊。怎么能体现中沙友谊？我们当时提出"海上丝绸之路"的理念，我们构思了一条船，通过收集资料发现沙特和中国早在明朝以前，就有非常密切的贸易往来，贸易往来的载体就是船。我们通过查各种资料，还专门找到北京大学阿拉伯语系付志明老师，让他跟我们讲中东历史和中国的联系，最后发现"海上丝绸之路"是一个线索。而它有形的东西就是"船"。"船"从沙特"漂浮"到上海就是这么一个概念。这个概念也引起了评委会的共鸣。这个船立意确定以后，空间体验和展陈布置怎么把船的空间用起来是关键。首先，我们把船架起来。因为世博会的展览时间从五月到十月是上海最热、最潮的季节。船架起来不光要体现出船的漂浮感，更要使用船体的投影来给排队等候的观众遮阳，通过架空船下的压力会加速风速，改善等候空间的微气候。同时我们在船的下面布置了一个游客等候的舞台，游客在船下面等候的时候可以观看预热的舞蹈。我们查了往届的世博会资料，发现世博会游客排队等候时间最长是三个半小时，所以我们按照三个小时左右等候的时间，来设计这个等候流线。但实际的等候时间大大超出，是我们没有预料到的。

我小时候看过童话书《一千零一夜》，给我印象特别深的是一个叫"阿拉伯魔毯"的小故事。我们就想能不能用"魔毯"的概念来设计，让观众有一种在魔毯上"漂浮"的体验。我们采用弧形的自动步道，架在这个船的内壳上来观看电影，颠覆了观众对传统观看电影的体验。我们取名叫"融入式"，现在叫"沉浸式"。我觉得这就是沙特馆受热捧最关键的一个因素。值得我们高兴的是：通过我们的思考，给大家创造了一个全新的参观体验。我觉得这就是最有价值的。普利兹克奖获得者葡萄牙的德莫拉有一句话：建筑师并不能发明什么，但是在解决整个问题的过程当中而产生的想法，就是最有价值的。所以对建筑创作来说，很少能有人发明了什么，但可以通过参观体验的创新比如像"全景立体融入式参观方式"，来体现建筑师工作的价值所在。

张力： 谈一下您认为的好建筑的标准，什么是中国好建筑？

王振军： 我刚从哈尔滨评优回来，看到很多优秀的建筑作品。在这个信息化时代，虽然由于网络的"同时效应"，导致信息爆炸，或者叫"泛中心论"从而没有绝对权威，但我觉得好建筑还是有"共识"的。我觉得好建筑应该具有以下特点：第一，建筑作品要有很高的感知度。自娱自乐、曲高和寡是不提倡的，肯定也不是好建筑。沙特馆的感知度就比较高，我们没有把外形做得很怪，而且它的功能都可以解释，游客参观后很喜欢，很认同，这就是感知度，因为建筑就是给老百姓做的、给使用方做的。我们黄总（黄星元）最近出了一本书叫《清新的建筑》，里面就说到：建筑不能做得太晦涩。第二，要充分地融入环境。这一点说起来容易，但做起来其实很难。首先甲方要求你要从环境中突围出来、彰显出来、有标志性。我们到欧洲历史比较长的国家去看，城市里突出的只是教堂、市政厅。当然现在不同的项目强调"在地性（in-site）"就是建筑要融入到这个环境里。我在《蔓·设计》那本书中提到：要发现场地里面的深层结构，然后融入它，有机地融合在一起。第三，形式和功能统一。建筑外部造型和里面的功能性都要有用。如果花很多钱去做一个建筑，纯粹是做个外形而忽略内部的功能，这不是建筑师该做的事。第四，空间体验感人。空间是建筑的主角，两百多年来都没有改变。比如到路易斯·康做的教堂，现在进去还会感动。建筑的室内外空间体验都很重要。我们现在的新机场项目，因为是总包，所以建筑、室内、景观，包括幕墙、照明等都是一起做的，当然我们会找一些专业公司配合一起来做，但是我们会全程主持把控室内外效果。第五、关照文化传承。中国是一个拥有五千多年历史的文明古国，中国建筑师不管做什么类型的建筑，当然特别是文化建筑类型，传承的责任会更大。说到文化传承，要说起沙特馆当时为什么会被选上？因为我们设计的船头偏了三十多度指向麦加。另外，参观的流线，是和麦加朝圣的流线方向一致的。后来与甲方签完合同后，他们的部长说："王先生，你对我们伊斯兰文化很有研究"。我们通过一个角度的偏转，包括流线的设计，最终得到甲方的认可，这就是因对文化传承的关注而取得用户认可的实例，也提高了沙特馆的观众感知度。所以文化传承很重要，中国现在国力增强以后，在一系列的重大工程里面都要体现出中国文化。特别是从一个大国变成强国的时候，就更要注重这方面的思考。我觉得中国建筑师的创作要更突显中国特色，这不应是表象的，而应是深层的。

张力： 谈一下您最近做的那个大项目？

王振军： 南京紫光集成电路项目是投资七百亿的一个工业建筑，建筑用地是50公顷，建筑面积将近60万m^2，是一个特别典型的高科技工业建筑。为什么叫高科技？因为南京紫光现在生产的是12英寸的存储器产品，是和美国在国际上处于同一个水平线上的产品。这个工业建筑的创作让我感慨很多，我觉得对于中国电子院这样一个典型传统的工业院来说，工业建筑是未来需要花力气去投入建筑创作的特别好的一个建筑类型。

南京紫光项目的特点：

1. 场面宏大。厂房本身长500m，宽210m，高34m，是一个巨型长方体。这个尺度在民用建筑里基本上是找不到的。

2. 对物理环境要求达到顶级。洁净、抗微振、消防的要求都非常高。

3. 建筑特别本质。一是从成本上不可能给建筑师更多的投资来做花拳绣腿的东西。二是工艺空间要求本质纯净，空间里不能有别的东西，否则会影响生产的洁净。

张力： 要突出它工业生产的本质特点？

王振军： 厂房部分本质上不能像民用建筑那样过多地修

饰和创作。成本上也只允许把重点放到生产空间的塑造及洁净空间、微振空间的营造上。其他部分的造价就被限制在一个非常狭小的范围内，不管是外墙材料还是内装材料，都不可能太有富余。但厂区里面除了有生产空间外，还有很多研究人员活动的空间。

现在甲方、投资商，对厂区的研发、建筑室内外环境要求非常高，比如，门厅以前只要求能组织好人流，能迅速地疏散。而现在要求在门厅展示高科技产品，还要把企业文化展示出来，就要有很好的空间体验。因为投资这么大的精密产品生产，无论是研发人员，工作人员，还是工人们心理压力都很大，所以投资方就对建筑师提出要求，要给这些工作人员塑造更高品质的、不同于一般传统工业厂区的工作环境。

张力： 南京紫光这类建筑的设计难点在哪里？

王振军： 设计难点就是只能在完全配合工艺流程的前提下进行创作。建筑师的价值就是怎么把自己的创作和工艺的要求、公用设备的要求非常紧密地结合在一起并有所创新。一般这种项目我们都是双总师，一个工艺总师，一个是建筑总师，这两个总师都要密切配合。但有时候这些限制会给创作带来很多奇迹。比如南京紫光，它有一个与这个厂房平行的研发楼。一般的研发楼开窗要求均质、大小差不多。这个研发楼总共是五层，底下主要是测试，还有一些数据中心，不需要开窗或者开很小窗就行。四、五层是研发楼，要求开大窗。我们就把开窗做了一个退晕，从下面不开窗到开小窗，然后上面再开大窗，这样从下往上就可以做成渐变，建筑外立面变得更丰富。所以很多人说：你这窗户开成这样挺好看的，我说：我这是工艺要求、功能要求。这就是创作有意思的地方，其实有时候没有限制反而不好做。

北京按摩医院是中国残联的项目，他们想把这个项目做成一个无障碍设计的范例，要在全国医院里推广。我和中国残联的吕世民副主席一直在交流，要把它做成一个国际上标准无障碍的设计。因为按摩院的医生很多是弱视，只能看到一点光，所以我们要设身处地去考虑他们的体验，这就要在设计中有爱心、耐心、细心、匠心，我们叫四"心"。无障碍设计虽然是很复杂的一套系统。但是我觉得只要用心去做就能做好。

张力： 请您谈一下您的创作理念和"蔓·设计"？

王振军： 工作室至今已经成立整整十年了，最近也一直在总结这些创作理念。我出《蔓·设计》那本书其实也是在总结。我认为：第一，中国建筑师要有自己独立的思考。因为，我觉得中国建筑从20世纪70年代末敞开大门开放，国外事务所的各种思潮、各种理念都引进来了。随着国家综合实力的增强，中国建筑师的分量会越来越重，就必须要有自己独立的思考。如果你没有思考、没有理念，那你就是不同风格的复印机，人家做什么你抄过来弄一弄，新鲜几天，时尚劲很快就过去了，这肯定是不行的。第二，中国建筑师要有自己的创作哲学。现在由于各种原因不管各个城市从官员到设计人员，特别是年轻建筑师都很浮躁，这种氛围使建筑师正在经受一个非常严酷的考验。第三，不断总结设计理念并告诉年轻建筑师。比如我的工作室，年轻建筑师一拨拨进来，怎么去带他们？我在学校教书的时候就有一种说法："建筑学只能学没法教"。从传统的角度来说建筑学确实没有一套特别完善的体系告诉你该怎么做。因为建筑学带有艺术的成分，艺术是没法教的，它要不断地创新，不断地制造陌生感。我也和很多高校老师交流，他们把这个叫"解释焦虑症"，因为建筑学没有标准答案。因为我在工作室也带年轻人，包括带研究生，这个建筑为什么很好，那个建筑为什么不是很好，每个人都得有自己的价值观，但实际上还是有很多共识的，所以我就把这些年的思考整理到《蔓·设计》那本书里。"蔓·设计"至少是两个维度：第一个维度：拿到一个项目以后要寻找这个场地的"自然场力"。每一块场地都有它自然留存的东西。比如观察一块石头肌理纹理、一条溪流和它的流速快慢，这些都和"自然场力"有关系。如果处在一个比较开阔的地方，常年被风雨洗刷以后，石头就很光滑。这就是"自然场力"造就下来的。比如财源中心沿着长安街将近280m长的地块，四周的建筑都盖好了，南边是建外SOHO，东边是环球中心，西边是一个公寓，每一栋建筑都有它自身的特点和风格及对左邻右舍的要求。例如公寓朝东有窗户，新建建筑如果离公寓东边太近就会遮挡它，如果遮挡到它就是破坏了这个"场力"的结构。这个结构有很多含义在里面：深层结构、空间结构、材料、日照、消防等。

综合所有因素最后建设出来的建筑叫建构，水、暖、电非常完美地组合在一起叫建构。中国现在的好建筑越来越多，也越来越讲究，就是建构出来的作品，要像自然界一样"有机"。我为什么说像自然界一样有机？因为中国历史几千年以来，再好的建筑，评价再高的建筑，都没有自然界呈现出来的状态优美。这需要我们深思。所以我们要向自然学习，但不是简单模仿。第二个维度：在以上的哲学认知下，要把这个认知贯穿到设计的全过程，或蔓延执行到项目的全过程。从项目建议书、可研、概念设计、建筑设计、初步设计、景观设计、室内设计、施工图设计、工地服务，要全过程蔓延。当时我写《蔓·设计》这本书想了好几个"màn"。在新华字典里：1，"曼"是点式的；2，"漫"是平面的；3，"蔓"它是立体的，立体的蔓延就是这个"蔓"。所以我们就选择这个"蔓"。可能大家也会有疑惑，都按这个"共识"来创作，那最后做出来的作品是不是都一样了？这个不必担心，我们强调艺术想象+理性分析。因为你遇到的项目在不同的地点、不同的城市、不同的地域，有不同的甲方，创作出来的作品肯定是不一样的，因为你是分析出来的，并不是表象照搬过去的。所以建筑作品虽然是在同一个"共识"下来创作的，但它呈现的状态是不一样。

张力： 前段时间说中国是国际大腕的试验场，中国的建筑市场乱象您怎么看？

王振军： 我觉得甲方很多项目选择外国建筑师，说明中国建筑师做得还不够精彩。中国建筑设计师还处在一个赶超的阶段，西方从维特鲁威开始快两百年了，从柯布西耶开始也近一百年了，但中国建筑才短短30年，虽然我们在二十世纪三十年代、四十年代，杨廷宝、刘敦桢、梁思成这一代建筑师是跟国际接轨的，但中间中断了。这几年咱们自己原创的好建筑作品越来越多，可以预见，我们和国际建筑师的差距整体上会越来越小。

张力： 中国建筑学的希望在哪里？

王振军： 我最近也参加一些研讨会，看到一些建筑师已经创立自己的认知体系，像崔愷院士在他的"本土设计"里就建立了很有高度和广度的哲学体系。还有像孟院士等很多建筑师群体都在建立自己的认知体系。在一定的共识下，建立自己不同的体系，这就是未来建筑创新的潜力所在。日本的建筑师每个人都有一套自己的东西，每年都出一本书，做出来的作品也都不一样，精彩纷呈。我觉得中国建筑师、包括管理者下一步都需要提倡、鼓励建筑师建立自己的认知体系，这样每个人做出来作品都会有自己的特点。正所谓"和而不同、同则不继"，中国建筑的未来值得期待。

AT 灵魂三问"蔓·设计"与"大建筑观"

一. 这几年在忙什么：

由于工业院出身的缘故，大的小的、工业的民用的什么都在做，有芯片和 TFT 厂房、数据中心、机场塔台，也有博物馆、科研建筑、总部大楼、商业综合体甚至 200 多 m² 的城市驿站（公共卫生间）等，还有一些就是集团技委会、建筑专委会的一部分技术管理、科研管理等工作。

除了项目，工作室最近在进行总结，从 2009 年成立至今已经 13 年，原来想在 2019 年底做的回顾展也因突发卫生公共事件一直拖到现在还未实现，所以想是否先出个册子纪念一下，这样对老同志是个交代，对新同志是个引导。

一句话：真诚的设计，真诚的生活，所以很忙、很累也很充实。说实话，有一定的成就感，但缺乏与之匹配的幸福感。

二. 最近有哪些思考什么：

问题 1、"大建筑观"： 作为工业院出身的设计院的建筑师，对建筑类型不能有成见，不管什么类型，建筑创作的本质都是一样的，只是这些本质要素相互之间的比例和先后关系略有差异而已。这从投入比例就能说明问题，典型的例子比如博物馆土建的投入可能要占到 50% 以上，但对于芯片厂土建的投资也就是 10% 左右，但要说明的是：一个大型芯片工厂的投资不亚于一个机场的投资，即使土建仅占 10%，他的单方土建造价绝对数还是相当可观的，加上它特有的超大尺度，所以建筑师还是很有发挥余地的。

我最近一直试图用"大建筑观"来说服我们院的管理层，摒弃掉以前工业建筑和民用建筑简单的二分法，这样就可以为建筑师创造一个更加宽阔的工作氛围和平台，我列出的理由如下：

1. 溯源西方近 150 年的近现代建筑发展史，从一开始就没有将工业建筑和民用建筑概念分开。现代建筑先驱 P. 贝伦斯第一个现代建筑开山之作就是工业建筑——柏林通用电气（AEG）透平机车间，他的门徒、也就是现代建筑大师格罗皮乌斯、密斯、柯布西埃也都是以工业建筑作品开启了自己的职业生涯，纵横于工业建筑和民用建筑领域。

2. 进入当下信息时代或者说智能化社会，新的建筑类型不断涌现，无疑是在印证这种绝对的、生硬的工业建筑与民用建筑分类已不合适。

3. 从学术上讲，本质上所有建筑类型都有工艺设计存在，芯片厂、医院、酒店、博物馆、数据中心都有工艺，仅仅是功能类型不同而已，对空间体验、形式美及其与功能的平衡要求程度不同而已。从建筑学的本质来讲都是一样的。

4. 目前设计市场处在粥少僧多的状态，设计院处于被选择的地位，所以市场决定了：设计院应该也必须保持一个开放的心态，有什么就做什么。现实也要求我们市场有什么就能做什么。市场决定了你为了生存只能模糊分类。

5. 针对我国已经从解决"有没有"进入"好不好"的发展阶段，业主对工业建筑特别是高科技工业建筑的总体要求包括空间环境质量、艺术性等已经不亚于民用建筑，南京台积电、江苏淮安实联化工厂（西扎事务所）就是明证。

6. 从我们的老客户自身的发展来看，原来很多搞工业的客户现在开始转型做非工业项目，如果要把这种传统的信任延续下去，我们必须是多面手。

7. 在上述这种条件下，仍要刻意强调工业建筑和民用建筑之分，无疑是作茧自缚，特别是本来就不强大的建筑师队伍还要截然分开，变的都不强大，到头来两头都做不好。攥成一个拳头打出去才会有力量。

8. 遵循建筑学基本要素要求及其外延要求，是所有建筑类型创作中建筑师必须遵循的基本职业准则，而类型只是具体设计内容不同而已。摒弃惯性的、简单化的二元分类做法，中国电子院定会抓住当前的市场契机，迎来更新更大的发展。

问题 2、团队的共识与创新性： 作为一个团队，如果没有一定的共识效率就会很低。所以找到一个大伙都坚信的东西，然后坚持追求下去很重要，工作室通过长期的摸索、总结，我们把对建筑的理解整理成做"蔓·设计"理论。随着工作室创作理念——"蔓·设计"思想的逐渐成形并成为共识，也随着一批骨干的涌现，我感觉工作室的创作已走在比较正确的方向上，这个时候如果对建筑语言、手法的研究跟不上的话，创作上又很难出现百花齐放的局面，这是搞艺术的大忌，这又与建筑师的追求相违背，所以我们准备今年举行一系列的研讨会，针对建筑语言、手法的变化来研讨。创作思想的统一加上手法语言的多元，我想工作室一定会产生出很多好的设计、好的作品。

我们探究创新的三大驱动力：一是人的审美疲劳，二是环境的改变，三是建筑师的自我批判。我们现在希望通过对手法的探讨激发团队创新能力。"人类到目前为止所取得的成就，都是在批判性思维下取得的"。从这个意义上来说，作为工作室共识的"蔓·设计"理论也是可以拿来批判或者说补充的，这样团队才会有生命力。

我们希望这个团队在对学术的思考上具备一定的深度、在作品的表达上具备一定的广度，假以时日一定能为设计院培养出一批批建筑业务骨干。

问题 3、最近还在一直思考节能、绿色和低碳三者间的关系，怎么把大势所趋与"蔓·设计"理念完全融到一起，现在总结出来就是：节能是国家政策要求、绿色是鼓励和引导措施，低碳是前两者活动的结果。节能是因，低碳是果。那么下一步就是如何定量地来解决或者说表达这几个方面问题。感觉这个问题很复杂，还在研究过程中。

我在想如果把这对三者的理解与"蔓·设计"融在一起来进行设计工作，就会使日常的设计工作很高效，也能够使"蔓·设计"之路走得更加坚实。

三. 上次出境旅游是啥时候？

两年前，突发卫生公共事件使工作室很多出国考察计划都搁浅了。

我觉得：建筑学的主观经验与其他艺术最大的区别就在于：建筑学必须经以人的身体作为尺度，而且建筑从根本上就是构筑我们可体验的的物质容器，所以建筑必须要去体验才能辨别其优劣。所谓创作中的灵光乍现或者妙手偶得，其实是建筑师对世界万物、人间百态后面隐含的规律和关系的深刻洞察后而激发出的，可遇不可求，并且需要长期学识和经验的积累，从本质上说也是建筑师理性思考的结果。这就是建筑实地考察的价值所在。

清华大学一位学者提出的"信息茧房"的概念：就是说互联网使人类族群极度分化，每个人都被自己所感兴趣的信息所围绕，信息反而越来越固化。那么一个团队一起来品鉴讨论一些经典或者有争议的建筑，对大家拓展自我、达成共识很有好处。

工作室之前一直每年会组织大家在世界经典建筑、新建筑中挑选一些去体验，希望突发卫生公共事件尽快结束，能恢复这一很有价值的建筑考察活动。

"蔓·设计"中的蔓访谈

2023年9月19日，王振军工作室+"蔓·设计"研究中心十五年作品展开幕式及学术沙龙在北京举行，围绕"生长与蔓延"与诸多院士、大师、各大设计院总建筑师、高校学者以及行业同仁们，对其十五年的建筑实践进行了深入的交流和探讨。11月2日，《城市 环境 设计》（UED）杂志新媒体负责人、内容主编孙宁卿来到展览现场，与王振军总建筑师面对面，展开了一场围绕"蔓·设计"的"蔓"对话。

"蔓·设计"的诞生

孙宁卿："蔓·设计"是您工作室的设计思想核心，您也有专著围绕这个议题。在此能否请您先为我们的读者简单介绍一下"蔓·设计"的由来和它的具体内容？

王振军："蔓·设计"诞生的背景有两个主要因素。

工作室在2009年成立时加上我一共是6个人，到现在发展为30个人，是涵盖建筑设计全过程、景观、室内以及产品设计的综合设计中心。其实发展和扩张并不算快，应该算比较保守的。

大概在2013、2014年左右，因为业务的拓展，工作室突然增加了10多个新人，明显感觉到工作效率降低。因为我也是从年轻的时候过来的，刚毕业的年轻人对建筑很难有系统的认识。所以，为了在创作时不纠结、不跑偏，提高效率，当时工作室非常需要有一个"共识"来告诉大家，给年轻人"讲明白"，这样无论是设计还是管理的效率都有所提高。

另一个背景就是2016年是我从母校湖南大学本科毕业30周年，我也是想在这个节点把自己的设计思想再梳理一遍。于是就有了"蔓·设计"的诞生契机。

总的来说，我觉得做建筑其实就是两件事，一是你怎么看待建筑，另一个就是你怎么表达它。"蔓·设计"理论正是回答了这两个问题。

比如说扎哈和奇普菲尔德，他们都毕业于伦敦建筑联盟学院（AA），但是他们对建筑的表达，完全是两种态度和方法，这就是如何看待建筑和如何表达建筑的差别。

我们首先梳理分析了建筑学科的核心要素的构成，将之分解为建筑自治性范畴的元素，场地、空间、材料等；建筑的外延，如绿色建筑、装配式、人文、社会学等问题；最表层的是建筑的表达，例如风格、手法等。

然后就延伸到"蔓·设计"的具体内容，首先从哲学角度是把建筑包括人都看成自然界有机物态的一种，建筑在介入环境中时，它是后来者，应和自然界各种有机物产生关系。这种认知基本上就表明了我们怎么看待建筑。而"蔓·设计"的方法论则是说如何去表达建筑。

我想从"生长与蔓延"两个维度来具体解读一下"蔓·设计"。

"生长"是哲学维度的，当建筑介入环境中时，我们要去寻找场地已经存在的自然场力，发掘场地之中存在的深层结构而不是表象，让建构的建筑呈现出来的状态像自然界一样有机。"蔓延"则是方法论维度，把上述设计哲学蔓延到建筑建构的全过程，打造高完成度的建筑作品。在"生长"上，提出了自然场力、深层结构等思考层面。例如自然场力包括阳光、水、土、风，也包括人、建成文脉等；深层结构包括宗教、价值观、结构的持力层，甚至古墓等各种隐含的元素，需要去挖掘。

而"蔓延"主要是从方法论角度出发。由于我们是工业院出身，所以接触到的建筑类型会比较多。我们希望将理念能蔓延到全过程、全类型，采用更多元的方法去实现有机建构的目标。

"生长"和"蔓延"的交集构成了这样一个核心，就是"有机建构"。它既是个动词，也是个名词，有机建构出来的作品，自然就符合我们认为是"好建筑"的属性，那就是要有良好的在地性，形式与内容统一，空间体验感人，具有可持续性，关照文化传承，有较高的感知度、突出的创新性等。

我提出感知度是因为建筑毕竟不是纯艺术品，它是要被人接受和使用的，不能太晦涩。

孙宁卿：后来效果怎么样？

王振军：通过捋清这些问题，团队达成了共识，解决了团队的"理论焦虑"，效率也就提高了很多。工作室成立至今15年，我们大概做了150多个项目，建成的有110多个。我们通过竞赛和各种建成作品的获奖，逐渐在增强团队自信心的同时，让同行和业主对我们团队有了更深入的了解。

孙宁卿： 接着我想问这种"共识"会不会束缚创作的多样性？

王振军： 我的回答是"不会"，因为"蔓·设计"只是一个基本的认识和表达，创新的潜力还有待于在"手法可行性的探索""新切入点的寻找"和"艺术想象力的激发"等方面进行实践。

另外也要感谢中国电子院，由于所处行业的特点，我们能够接触到世界最先进的工程技术和新材料、新标准。比如说常规的设计院做给水排水，可能就只能接触到 5、6 种水系统，而我们院经常一做厂房，都是 20 ~ 30 种水系统。另外，这个大平台也使我们这些建筑学毕业生对项目的类型没有了成见，让我们接触到了形形色色的项目，这个平台赋予了团队特别强的适应性和韧性，从而也使我们的作品呈现出多样化。

空港建设中的生长与蔓延

孙宁卿： 空港类建筑是你们工作室的一大特色，我看到放在了展览的开篇位置。

王振军： 空港项目确实是一大板块，我们在 2015 年到 2017 年之间，在大兴机场中标了 10 个项目，另外还有 2 个委托项目，一共建成了 12 个项目。

孙宁卿： 我看到工作室做了不少航空导航设施、空港配套建筑，比如你们曾发表在 UED 的北京大兴国际机场信息中心和指挥中心是非常经典的。

王振军： 机场指挥和信息中心（AOC+ITC）由 A 级数据中心、亚洲最大指挥大厅和信息研发办公楼等空间组成，承担着北京大兴国际机场的信息中心、通信枢纽中心、运行指挥中心三大功能。在设计上，我们通过分析功能需求、场地自然场力和深层结构，借用了"信息链"的概念，将指挥中心（AOC）和信息中心（ITC）及其数据中心连接在一起，并将不同尺度的自然庭院穿插其中。这一设计不仅强调了高智能建筑崇尚自然的理念，还表达出我们对"以人为本"这一设计原则的不懈追求。建筑体量被适度拉开，通过水平延展的通廊彼此连接，既梳理出场地轴线，又建立起建筑与机场景观的关联。整个建筑群的体量和空间重心适当上移，与机场高架的视觉效果相呼应。这个项目做完，外地很多甲方来参观后，就直接找到了我们。比如我们新建完成的烟台蓬莱国际机场信息中心工程项目。做完这个项目我们又中标了大兴机场的生活服务设施项目，投标时我们的概念叫"漂浮的阳光之城"。当时这个投标其实有点冒险，因为它周边盖完的很多的公寓都是传统阵列式的，一半朝南，一半朝北。阵列式的公寓就会有一半照不到阳光，其实在北京是不合适的。同时再考虑到这个楼和楼之间的日照间距，建筑会把用地占得很满。我觉得这是不合逻辑的，所以我们就做了一个"户户有阳光"的方案，把南北这两条都做成单廊，全南向。把东西向体量的做成内廊，这样就保证每一个房间都能够有阳光。

东西两侧再通过小阳台和深窗套来遮阳。公共的东西放在裙房，用一些地景式的绿色建筑手法，让人可以走上去。这种处理方式比"阵列式"的省出很多用地，现在的方案空出了很多空间作为运动场地。再比如在之后的住户分配时，几乎没有纠纷，因为每套房都有阳光，后来我们也因此获得了甲方一些直接委托的公寓项目。

孙宁卿： 你们设计塔台建筑是非常有经验的。

王振军： 我们先后做了首都机场 T2、T3 的塔台。其实，这些塔台是很多同行不愿意做的项目，一是因为技术难度高，二是因为塔台属于每个城市的门户建筑，审批过程严苛，往往方案报审还需要上高级别的专家评审会，比如在北京要上首艺委会等。

塔台的功能形制特点是下面和上面大，中间就是交通空间和管井，所以形式的逻辑基本如此。T2 塔台的构思是天坛的形状，以呼应北京古都的元素。塔台要做出特点是比较难的，因为它的功能单加上国际民航标准限制，基本决定了这类建筑很难有所突破。

做 T3 塔台的时候，本来福斯特（Norman Foster）也想参与，但是出于保密原因没做成。我们中标的主要原因是我们提出了 T3 塔台与航站楼"互为投影"的概念，它的形式呼应了 T3 航站楼的轮廓，就像是把航站楼的轮廓竖起来，另外意象形状像古代喝酒的"觚"，因为机场就是迎来送往的一个场合，表达一种"礼节"的概念。所以把这两个概念合到一起，就被选中了，定方案的时候福斯特也来了，他也很喜欢这个概念。最后，这个项目建成后获得了国家优质设计银奖和中国建筑学会60年创作大奖等奖项。

我们后来又做了西安塔台，用的是大雁塔、小雁塔的密檐塔的概念。郑州机场塔台是使用了河南省博物馆的镇馆之宝——贾湖骨笛（编者注：也有称贾湖骨龠）的意象作为设计概念。之后还陆续做了天津、珠海等机场塔台。

因为T2、T3塔台具有很高的知名度，我们工作室因为这类建筑有了些名气，类似的甲方就会陆续找来。后来我们做了许多塔台建筑，也主持编制了相关的规范。

作为里程碑与起点的沙特馆

孙宁卿： 在您2017年出版的专著作品集《蔓·设计》一书中，沙特馆被放在项目部分的第一位，我有注意到，在这次展览的开幕式演讲中，您又把沙特馆放在了典范项目介绍中的最后一个。所以无论是起点还是终点，我觉得世博会沙特馆的项目对您都是一个里程碑，是这样吗？

王振军： 可以这么讲，这个项目让我们收获很多。

上海世博会沙特馆有两个特点。第一它是世博会上42个外国馆中，唯一在全球通过竞赛征集方案和唯一一个由中国人原创完成全部设计的。其余41个外国馆的方案都是参展国本国建筑师设计的，然后我们本地的设计院配合去做施工图。英国馆、德国馆等都是这样完成的。

第二个特点是它就像是一匹黑马，一开始没人关注，甚至我们的中标消息也只有《世界建筑》刊登了一个很小的像豆腐块大的新闻，这个过程中，中外媒体将重点全部投向了英国馆等欧洲国家馆。

当时沙特馆筹建负责人利雅得大学建筑系的教授阿罕姆迪总是发邮件问我，未来在世博会沙特馆能排第几，所以我压力也很大，在开馆前中外媒体仅从对建筑外观的判断去追捧诸如英国馆、德国馆之类的外形酷炫的建筑，其实这些馆内部的状态大家都无从知晓。

这个项目要从2007年12月说起，沙特馆通过自己的建筑事务部向全球发布了一个征集消息。它的项目征集时是没有门槛的，文本要求也很简单，不少于五张A4纸就行，当然这一轮征集也没有保底费。任务书也不复杂，只有两个要求，第一要展现中沙友谊，第二要反映沙特的美好。

我们从朋友那得知了这个消息，抱着放松的心态和全力以赴的态度去做的。结果我们的方案就选上了。当时第一轮从全球收集到46个有效方案，然后里面选出来5个做第二轮。我们有2个方案进入了第二轮，另外还做了一个以"阿拉伯魔方"为概念的方案，排在第四。

这个项目做的过程也比较低效，主要原因是双方要远程配合，再加上沙特方假期多，绝对工作时长比较短，贻误了很多关键节点，导致我们还有很多好想法没能实现。

孙宁卿： 确实，我记得当时沙特馆确实不像英国馆那样简直变成了一个媒体现象。

王振军： 对，但是开幕之后，沙特馆的人流剧增，而发生了拥挤事故，马上引发了中外媒体关注，所以沙特馆是由观众把媒体的报道带动起来的，后来沙特馆最长的排队时间达到了9个小时。而我们是按照预测2~3小时的排队时长设计的，以致于完全超出设计负荷，好在我们的参观流线是单向不可逆的，比较容易控制节奏。

之前因为只看到各个馆的外观，无法入内参观体验，我也认为英国馆、德国馆会比较受欢迎。其实从开幕看到最后，我发现建筑是否受欢迎，观众会作出选择，你要论酷炫的话，沙特馆肯定不算酷炫，它就是一个传统的"船"的概念。而且它的完成度在我看来并不算非常满意。例如，我们没想"船"下做那么多柱子，但是因为上海的地质等不利因素，导致结构计算上柱子减不下来。

通过沙特馆项目建设的全过程，我们对什么是"好建筑"有了更清晰的认识，首先它要具有一个较高的感知度，另一个要看空间体验是否感人，第三个要有创新的手法。

比如有些外国馆虽然外形做得很酷炫，但是进入它的空间走一圈，出来就忘掉了，没什么记忆点。

后来很多记者也问，为什么沙特馆能这么火，我觉得是因为我们的创意为观众提供了一种全新的看电影的方式和空间体验。我们把展馆做成一个"船"形，然后利用船内壳做了屏幕。展览部分使用了26个投影仪从吊顶打到船的内壳，用投影铺满，铺完以后做自动人行步道把观众带入其中。这个灵感就来自"阿拉伯魔毯"的故事，我们用一种颠覆传统空间设计的手法，将观众带入了历史、现实、神话交替运转的梦幻之地。我们用了一个全球独创的"全景立体融入式"的参观方式，为中外观众打造了一场更逼真、更融入、更震撼的参观体验。

孙宁卿： 沙特馆最大的一个特点是它创设的一个多维的观看场所，这其实是由于弧形步道技术的发展，以及把现实和虚拟进行交融，创造了新的场所。当年是通过电子媒介和机械技术，那现在全新的虚拟与现实融合的空间创新，会给建筑学带来什么样的转变吗？

王振军： 沙特馆其实没有用特别尖端的技术，自动步道很多国产厂家可以生产，但无非是弧形自动步道要先进、复杂一些，可能会涉及更多一些专利技术。另外，也有记者以为我们用了什么3d MAX技术，实际上只是多几个投影仪同时投屏而已。

我觉得不应太夸大技术方面，沙特馆是靠建筑师的创意才改变了这种传统的参观和展陈方式。而技术方面本质上只是对创新的一种助力。

沙特馆之所以对我而言有里程碑意义是因为这一经历更使我厘清"什么是好建筑"这个问题。沙特馆从无到有的过程，正好是我们进行反思和学习的过程。另外，沙特馆项目采用的是欧洲的建筑师负责制，建筑师要控制和参与所有的材料、设备、系统的确定和采购。这对我而言是从无到有的实践经验。这个项目经历使我们建立了比较好的职业信心，同时在理论上也更加自信了。

大建筑观

孙宁卿： 您强调"大建筑观"——凡是满足人类活动需要的物理和精神空间的建构皆为建筑，在如此宽泛的定义中，建筑师究竟扮演了怎样的角色？

王振军： 首先要明晰，我说的"大建筑观"不是"广义建筑学"。

这个"大建筑观"，其实是针对"工业民用二元分法"对我们院在面对高科技工业建筑和工厂建筑创作时所造成的束缚而提出的。我经常参加全国的各类评优，比如数据中心建筑，可能有时会被分到民用建筑组，有时又会被分到工业建筑组，而分到不同类别里，评价标准就会不同。这一现象说明原来的"二分法"已不能适应信息时代的需求，我们必须反思。现在很多工业建筑对美学要求很高的，不少新产业的老总都是留美、留欧回来的，对工业美学是很讲究的。如果建筑师主观认为民用建筑应该好好做，工业建筑就简单做，那会产生很大问题。

再拿彼得·贝伦斯（Peter Behrens）的透平机车间举例，它是1921年设计的，已经100多年了，我想贝伦斯设计的时候肯定并没有把它设定成是工业建筑还是民用建筑。

孙宁卿： 你们算是做过这个时代最复杂的项目。

王振军： 对。所以后来我们就提出"大建筑观"。比如大家经常说彼得·贝伦斯的透平机车间，我经常跟团队建筑师说，建筑史上第一个现代主义建筑其实是工业建筑，另外也会给管理层举例说明，第一座现代意义上的知名工业建筑其实是建筑师主持完成的。

孙宁卿： 当时德意志制造联盟里很多建筑师，从工业建筑和纯功能性建筑中也反映了现代建筑本身就有一种类似你们的"大建筑观"的特质。

王振军： 所以在这次展览序言最后引用了美国哲学家约翰·杜威（John Dewey）所提出的一个观点：哲学是始于惊奇，终于理解，但是艺术——也包括建筑——是始于理解，终于惊奇！理解建筑、研究建筑，然后创作建筑，应该就是我们职业建筑师基本的工作路径所在。

孙宁卿： 对，因为如果德国人当时只是把建筑想象成一种工业构筑物，想象成一种可以快速制造的产品的话，那就不会有后面的德国制造，不会有德国设计。

王振军： 像100年前德国通用电气这样一个工厂，在那么久之前就设有总建筑师的角色，这个概念挺先进的。咱们的设计院到现在在职称系统中都是工程师，没有建筑师之说。不管怎样，现在至少在工作室层面，我们不会把工业建筑、民用建筑去做严格区分，都一视同仁。

比如大兴机场的车辆维修中心项目，一般人或许就做了个大车库，我们根据车的尺度，根据维修照度需要，把屋顶掀起一些来做采光，用被动式绿色建筑的手法，节约了照明用电。这个项目的甲方说我们把车库都做成博物馆了，我跟甲方说，我们现在把所有建筑都是当博物馆做的。因为整体来说，这些建筑都有"工艺"——各种流线、人流、物流。车辆维修有它的工艺，坏的车从哪进，好的车从哪出来，这就是工艺。博物馆的展品，观众如何进出，它的工艺性不比维修站简单，民用建筑里的医院则更加复杂。所以不能再按有无工艺来简单区分工业建筑与民用建筑了。另外还有一个背景，现在尤其是电子厂、生物制药厂，全是超洁净空间，对色彩和材料以及环境品质都有更高的要求，比一般民用建筑的要求高很多，所以我的"大建筑观"是针对这样的现实而提出的。

孙宁卿： 现在工业建筑的设计院领域能跳出这种分类方式的多吗？

王振军： 肯定不多。这种惯性和设计文化还是比较根深蒂固的，当然还有其他一些现实原因，需要很长的时间来调整思想。至少我已经在身体力行了。

孙宁卿： 紫光南京集成电路基地项目（一期）给人的印象是非常漂亮的，在你们作品中曝光率也很高，这个项目有什么特点？

王振军： 其实这个厂的投资堪比北京大兴国际机场。用地有 56 公顷，建筑面积达到 76 万 m^2，投资高达 750 亿。工艺设计刚开始着手的时候我们没怎么介入，厂区有多达 26 个栋号，被布置得特别复杂。后来通过研究芯片工厂的工艺逻辑，我发现在这复杂的表象背后，就是它们之间所呈现的"平行关系"，然后再用最短的连廊将它们串连起来，就可以满足工艺的要求。

结果就是现在所看到的规划结构。接下来在单体设计时，我们通过寻找母体来整合每一个厂区。主厂房的尺度达到 450m×180m，包括各种层高的空间以及设备层，还有大跨度的洁净空间，主厂房在剖面形成 U 形母体原型，其他建筑就依照这个原型来不断地重复它。最后用一些弹性空间诸如餐厅、会议室等来把他们串起来，这样秩序感自然产生了。

立面上根据"形式追随性能"的原则进行设计。形式根据变化的内容而采用不同尺度的窗户与玻璃幕墙，自然就产生了富有变化的、刚柔并济的效果，加上超长的尺度，配合简练的景观，表现出了独具特点的空间形象和品质。甲方对此也十分满意。

城市的建筑

孙宁卿： 上海世博会的主题是"城市，让生活更美好"，我们知道"蔓·设计"思想的成型虽然要晚于沙特馆建成，但现在是否可以谈谈您怎么看待或处理设计中建筑与城市的关系。

王振军： 从"蔓·设计"说起，我始终认为不管一个单体也好，还是一个组团、园区也好，它们都是自然中的一个有机物态。任何建筑，哪怕是城市，尺度再大也大不过大自然，就像凯文·罗奇（Kevin Roche）所说"自然才具有最高的复杂性"，所以建筑一定要小心介入环境和照顾好与邻右舍的关系。这个结论在我们做了财源中心（IFC）和中央美术学院燕郊教学楼之后，有了更深的感受。

位于北京 CBD 的财源中心周围的建筑物都是各种非常规手法的、千姿百态的造型，这就是它所处的文脉，通过反复推敲，我们最后用了对比法和模块化的手法，用极简的方法来呼应周围的环境，加上深色玻璃幕墙，没有任何花拳绣腿的形式，现在看来它在 CBD 显得特别干净简练，自然也就突显了出来。事实证明我们当时的选择是正确的，由此看出，对城市的认识，也需要一个过程。

做中央美术学院燕郊教学楼时，周围被程式化的住宅围合，那它就需要根据自身的功能逻辑去有机地表达出来就可以了。

在新疆阿克苏三馆及市民服务中心和中国计算机博物馆的设计中，我们也遵循"蔓·设计"理念，赢得了评委的好评。

科技园区项目

孙宁卿： "大建筑观"下，很多建筑确实呈现了与众不同的质量。你们做的很多科技园区都不是传统工业区的感觉。

王振军： 是的，比如我们最早在上海做的浦东软件园一、二期工程，是我们国家最早的软件园。由于造价限制，我们就着意从园区空间风格塑造、被动式绿建等方面入手。园区的甲方是从普林斯顿大学回来的，他很喜欢这个学校的风格和环境。我们选择了不到2毛钱一块的陶土砖做立面，显然贴上去效果还不错。这个园区很有某种类似北美的大学校园的感觉。

孙宁卿：是的，这种建筑本身不需要太花哨的手法，光靠体量，它就像库哈斯写的 Bigness 一样，提出某一类建筑物——例如摩天大楼，光靠体量和尺度就达到了某种意义。

王振军：我们这两年还做了海盐的智能制造创新中心、中芯京城集成电路生产线等不少这类项目。最近刚完成北京中关村移动智能服务创新园，容积率达到了3.5，而且场地本身限高，所以密度也非常高。为了减小地上建筑的密度，我们将1/3的园区公共功能布置在地下。项目的甲方喜欢曼哈顿的那种小街区、独栋建筑的感觉，我们通过对不同的颜色、材料、肌理、细部的精心推敲和处理，用高完成度的有机建构为高密度科技园区的打造提供了样板。

孙宁卿：中关村移动智能服务创新园项目的体量精心适度，立面做得挺精致。建筑师通常处理这种表皮会想反映某种"诚实性"，会做成正交分格，但这个不是，看似某种透视线，但其实是平的，增加了一些可能是文丘里所说的矛盾性和复杂性。

王振军：是平的，但是这个耗材就会稍大一点。幸运的是这个项目的甲方不错，对品质很有追求。

孙宁卿：工业园区、高新技术园区其实随着发展和生产线的快速变化，会有灵活性的需求和固定的工艺之间的矛盾，您在实践过程中是怎么解决的？

王振军：对，肯定要有冗余的。设备系统，包括结构、荷载都要留出冗余。比如芯片生产，它的投资动辄几百个亿，它的生产线中工艺设备的资金占比可能要占到50%多，是非常昂贵的，所以所有的空间环境，包括随着它的制程越来越先进，要求越来越高，它的生产线可能不断地在改变，但厂房不太可能再随时扩建。所以要留出冗余和弹性空间。

一般来说在设计初始阶段主要由厂方科学家进行设计，提出工艺初始要求。以芯片厂举例，设计院的工艺师起着将科学家提出的条件进行转译的作用。然后工艺专业会给我们其他专业提条件，我们再用建筑专业的手段去实现。

孙宁卿：您觉得马斯克的超级工厂算不算建筑？因为马斯克是一个科幻迷式的工业主义者，他的超级工厂除了尺度以外，我觉得更像飞船、汽车一样的产品。

王振军：也算建筑，比如特斯拉在上海的工厂也是常规的设计院配合的。而且特斯拉的工厂，它不会比芯片厂更复杂，只不过是自动化率、机器人使用率比较高。特斯拉超级工厂系列呈现的是一种"高端制造"在建筑上的技术转移，把飞机、汽车制造等科技转移到了建筑中，是工业建筑向高品质发展的一种现象。

文化传承

孙宁卿：您对好建筑的判断中，刚才提到，除了"感知度高""空间体验感人"以外，还有很重要的一个点是"关照文化传承"，比如说几千年的中国历史，但文化传统本身非常难以在抽象与具象之间实现平衡。关于这一点，您在建筑设计中是如何考虑与权衡的呢？

王振军：一个比较有代表性的项目是已建成的中国残联北京按摩医院扩建项目，这也是一个竞赛项目，全国擅长做医院的很多设计院，包括中元、同济、哈工大设计院等都参加了。由于审批部门不同意坡屋顶，我们通过提取中国传统建筑的檐、梁、柱元素，转译并抽象成新的建筑语汇，植入到现在建筑体量中。我们通过"传承与提升"的设计概念，打造具有传统意蕴又有当代中国精神的建筑。同时在医院的空间序列设计上借鉴了中国建筑传统经典的处理手法，如檐下灰空间、月亮门、鱼池、迎客松、疗愈花园等。

尤其是这个建筑是中国残联的医院，很多医生是视弱人群，因此也是外交部指定的对外展示我国残疾人事业的窗口单位。另外对于体现传统建筑文化以及中医文化的氛围是有要求的。我们利用鱼池产生的流水声引导空间，主轴上做了疗愈花园等。另外，在造价不高的情况下，我们在建筑的西立面根据朝向做了深凹的外廊，用于遮阳而且能够放置花盆；我们也使用了下沉边庭等被动绿色建筑手法，将自然风和光引入到地下室和车库，使使用者完全感觉不到是在地下，这也形成了项目的一大亮点。

孙宁卿：最后，您对于工作室未来的发展计划是怎样的？

王振军：可能至少存在升级的问题，现在成立15年了，虽然一直走在健康的轨道上，但是随着中国基建形式的变化，还是要用"再上台阶"来迎接这种转变的，在思考的深度、创作的广度和作品的完成度上还有很多工作要做。而且现在行业僧多粥少，竞争会更加激烈，从增量时代到存量时代，可能项目类型也会有所变化，这都需要我们有所应对。

教育

孙宁卿："蔓·设计"理论最初有一个诞生动机就是因为工作室扩张很快，需要为年轻的入职建筑师解决理论焦虑。我知道您也有做一些教学工作，现在还在教课吗？

王振军：工作太忙，现在已经没有精力去上课了，主要通过带研究生的方式介入一些教学工作。

孙宁卿：之前您在湖南大学留校任教了两年，喜欢教设计课吗？

王振军：以前没有太喜欢。我现在想起来，因为本身自己对建筑的思考也不是很清晰，就还是这个理论焦虑并没有解决。另外当时我留校时实践经历还少，教学生也只是纸上谈兵，就觉得自己应该还是去做些项目以后再回来，现在随着自己通过实践总结、梳理清晰后，觉得可以带些年轻人，作为一个老建筑师，有责任把自己的一些体会告诉他们。我觉得建筑学是很难"教"的，可能在事务所里手把手教学生操作实际项目，更利于培养。

孙宁卿：对现在的建筑学和相关专业的同学，有没有什么指导性的建议？

王振军：我和本科生接触得少，硕士多一些。我觉得国外那种有实践经验的建筑师来参与教学会好很多。总的来说建筑学是一个实践性学科，但我觉得同时理论也很重要，如果理论焦虑不解决，不去看书和思考，建筑师永远都不可能走得太远。

我觉得那些建筑大师都是想明白了这两点。所以我觉得学生要通过多读书多思考，不要只看表象的东西，自己也要把建筑这个事想清楚。当然年轻的时候肯定就要带着这种焦虑成长——可能这就是年轻的一个特点。

孙宁卿：你喜欢的建筑师有哪些？

王振军：这个问题很难回答，现在不像年轻的时候特别崇拜某个建筑师了。

孙宁卿：以前有喜欢的建筑师？

王振军：当然有，比如安藤，我在1991年写了一篇文章《精神与形式——关于安藤忠雄与后现代主义》，登在《时代建筑》上。但是现在虽然觉得还是不错，但是没有那么崇拜了。还是一个从学习到仰望，再到理解，再实践，再总结，慢慢地自己有了理念，从而变得更坚定了。

孙宁卿：您后来还批评过一个现象说现在建筑作品的首要服务对象不再是客户，而变成了公众和社交媒体。建筑在这个社交媒体的时代又获得了前所未有的关注，但却不是以建筑本身——人们可能不再到建筑当中去看建筑，而更多是通过媒体去看建筑，尤其是网红建筑，这个现象您怎么看？

王振军：对，这个现象现在也挺严重的，可能和互联网的发展有关系，由于人们获得资讯的方式变得既容易又多元，所以评判的标准也变得特别含糊和混杂，很多媒体对事件的传播也让大众对什么是好的建筑产生了混淆。

孙宁卿：2019年到后来2022年12月之间的这段时间，您个人和工作室有什么变化吗？

王振军：感觉应该和其他设计院差不多。要说感受的话，其实最大的感叹还是刚才那句话：自然界才具有最高的复杂性。更印证了"蔓·设计"的哲学思路是对的，让我们认识到"人"只是大自然中的一个有机物，并没有什么特殊性，并不是说因为你能力强，那些细菌、病毒就饶过你。

孙宁卿：房地产的退潮对你们有影响吗？

王振军：其实展出的那么多项目里面大概只有一个商场是地产项目，工作室介入房地产少，受影响非常小。我觉得以后可能那些不做房地产而且创作能力强的建筑事务所会生存得很好，因为他们的创造力比较充足，设计深度又深，而且适应能力强。

孙宁卿：您是不是那种希望别人用一两个项目记住自己的建筑师？

王振军：我就是喜欢踏踏实实做些作品，并希望在用作品与前辈、同行交流的过程中学习提高，同时在这个过程中带领团队向前走。

王振军：生长与蔓延——《当代建筑》CA

CA: 请您和我们分享一下"蔓·设计"是如何得名的？

王振军： 在2016年，我迎来了从事建筑设计的第30个年头，我深感有必要对自己的建筑认知再进行一次严谨的梳理。此时，随着项目数量逐渐增加，工作室迎来了人员扩充的高峰，我有责任将我的建筑思考清晰准确地传达给这些年轻人。只有这样，我们才能在繁重的创作工作中保持思路清晰，避免纠结和偏离，高效、高质地让项目落地。

我提出"蔓·设计"理论是基于对建筑学的深入理解。建筑学涉及众多要素，理解其核心就要把握其构成要素：建筑学的自治性、建筑学的外延性及建筑的表现。通过对现实进行反思，加上从自然中得到的启示，我提出了向自然学习的理念。

在场地中，每一块砖、每一片石、每一株草、每一棵树都清晰地显示出了自然的秩序。这些秩序被精心组织和规划，展现出的每个细节都充满了建构之美，令人着迷并感动。

"蔓"这个字在辞海中的释义为五行属木，呈现为立体的蔓延。"蔓"本身就象征着自然界的蓬勃生机，非常贴切地表达了我们的设计追求。基于此，我提出了"蔓·设计"理论。

CA: "蔓·设计"理论回答了"怎么看待建筑"及"如何来表达建筑"这两个核心问题。请您结合实践案例，向我们介绍一下"蔓·设计"理论。

王振军： "蔓·设计"理论主张将人和建筑视为大自然生态的一部分，强调建筑师在介入自然环境时，应积极寻找项目场地中已存在的自然力量，并深入挖掘其中的深层结构。根据该理论，建筑的表达方式和最终呈现的效果应类似于自然界中的有机建构状态。这种深层结构包括业主的价值观及宗教信仰，也包括一些复杂工业厂房的技术逻辑等。例如：在中国残联按摩医院项目中，业主着重强调传统中医文化和传统建筑文化的重要性；在上海世博会沙特国家馆设计中，如果设计师不对沙特阿拉伯的宗教和文化进行深入研究，就无法顺利开展设计工作。在芯片工厂设计中，如果你不去深入研究学习芯片制程中的工艺要求，也无法开始厂区的整体设计。

CA: 我们了解到您工作室的核心追求是理性介入+有机建构。可否详细讲解一下您是如何在项目中落实这些工作方法和流程的？

王振军： "蔓·设计"理论主张任何建筑都应以理性的方式介入场地，促使所有参与其中的要素进行全过程的建构。在项目开始阶段，我们将对各种要素进行分解，根据它们对建筑的影响程度进行平衡和取舍。随后，我们将建筑的自主性和外延因素进行组合，并根据类型、地域、业主的价值观等因素，得出一种或数种表达方案，随后进行讨论和优化。我认为设计是一个逐步深入、蔓延的过程。

CA: 您长期从事电子工程、IT建筑、机场导航建筑、产业园区及博览馆等建筑的创作，主持的电子工业国家级重大项目为解决我国"缺芯少屏"问题做出了贡献。请您详细介绍一下您团队的创新研究。

王振军： 作为中国电子信息产业发展的重要支柱之一，中国电子院始终致力于推动关系国家产业安全的集成电路和新型显示器等产业的发展。在我院首创的"重大工业项目"双总师机制下，我本人作为建筑"总师"积极参与了多个国家级重大项目的设计工作，如武汉国家存储器基地项目（一期）、紫光南京集成电路基地项目、北京中芯京城（亦庄）项目、厦门天马T19项目等。在紫光南京集成电路基地项目中，我们通过发掘工艺设计中平行关系、寻找最短连接路径，提炼出建筑母体、形式及材料选择追随性能等手法，成功打造出一座具有现代高科技工业美学特征的集成电路生产园区。

CA: 您在进行机场导航建筑设计实践的同时，为行业标准编制做出了贡献。其中，您主持设计的亚洲最先进的北京大兴国际机场指挥和信息中心（AOC+ITC）成为我国机场智慧建筑的范例。请您详细介绍一下您团队在此类项目中的研究。

王振军： 北京大兴国际机场指挥和信息中心（AOC+ITC）由A级数据中心、亚洲最大指挥大厅和信息研发办公等空间组成，承担着北京大兴国际机场的信息中心、通信枢纽中心、运行指挥中心三大功能。在设计上，我们通过分析功能需求、场地自然场力和深层结构，借用了"信息链"的概念，将指挥中心（AOC）和信息中心（ITC）及其数据中心连接在一起，并将不同尺度的自然庭院穿插其中。这一设计不仅强调了高智能建筑崇尚自然的理念，还表达出我们对"以人为本"这一设计原则的不懈追求。建筑体量被适度拉开，通过水平延展的通廊彼此连接，既梳理出场地轴线，又建立起建筑与机场景观的关联。整个建筑群的体量和空间重心适当上移，与机场高架的视觉效果相呼应。

在A级数据中心机房部分的设计中，我们按照国际标准进行了适度超前设计，采用了8个机房模块一次施工、分部安装的方式，以便后期扩展和机房等级的提升。这种设计充分考虑了建筑全寿命期的节能和可持续发展。

该项目以数据中心和指挥大厅为平台，基于对北京大兴国际机场智能化信息的综合应用，集架构、系统、应用、管理及优化组合为一体，重点打造具有感知、传输、记忆、推理、判断和决策的智慧综合体。项目建成后，为机场工作人员提供了安全、高效、便利的工作场景，并为可持续发展的大兴智慧机场提供"最强大脑"。

CA：在我的印象中，中国电子院是以工业建筑为主营业务的大型设计院，但看您的展览基本感受不到工业建筑与民用建筑的明显区别，我想这应该和您对建筑的思考或个人经历有关，请您与大家分享一下。

王振军：一百年前，由P·贝伦斯设计的德国通用电气公司透平机车间诞生，其作为第一座"现代主义建筑"，向我们揭示了工业建筑和民用建筑在本质上的一致性。为了更好地反映这一理念，我们工作室在建筑创作中摒弃了传统的"二元划分法"，淡化了工业建筑和民用建筑的界限，提出了"大建筑观"。这一观念主张"像做博物馆一样做工业建筑"，强调在满足工艺需求的同时，注重建筑的艺术性和品质。

在我国进入新时期高质量发展阶段的背景下，我们遇到的业主认为对项目功能性和艺术性的追求同等重要。为了更好地满足这些需求，我们院在重要的工业项目中建立了"双总设计师制"。通过这一制度，建筑"总师"可以领衔完成工程设计全过程。在保证生产工艺需求的前提下，我们引入了"蔓·设计"理念和"技术、艺术一体化"的设计路线，塑造了技术与艺术相生共融、具有工业美学特点的现代高科技工厂样板。

这些样板项目不仅在功能上满足了生产需求，而且还展现了高科技工业建筑特有的美学特征。这一设计理念的实现，不仅提高了项目的品质和价值，而且为我们探索新时代下的建筑创作提供了新的思路和方法。

CA：在"蔓板行歌"宣传片中有一句话很打动我："工作室留下了一批有情怀的建筑师"。当下"行业下行""建筑师转行"，情怀与热爱显得尤为重要，请您结合个人从业经历，谈谈对建筑学、对情怀的理解。您作为工作室的主持人，是如何构建工作室培养体系的呢？换言之，您是如何让青年人愿意留在这个平台且保持创作热情的呢？

王振军：在过去的15年里，工作室通过学习、借鉴国际先进事务所的模式打造了这个建筑师平台，形成了"创作为上"的工作室文化，总结出了"蔓·设计"理念。在此共识下，我们成功地完成了许多获得业界认可的项目，并通过这些实践将一批年轻建筑学子培养成了建筑师。

在这里，我想强调的是，留住一批有情怀的建筑师，并使其保有创作热情，是以上各种因素综合在一起共同作用的结果。正如美国哲学家约翰·杜威（John Dewey）所说："艺术是始于理解，终于惊奇。"一位前辈也曾说过："搞艺术就是两件事，一是怎样看待这件事，另一个就是如何表达。""蔓·设计"理论正回答了"怎样看待建筑"，以及"如何表达建筑"这两个核心问题。

在"怎样看待建筑这件事"的问题上，我认为只有想清楚了，你的创作激情才会找到释放的地方。后续诸如"手法可能性的探索""新切入点的寻找""艺术想象力的激发"等方面，将成为我们未来无限的创新潜能的关键。

建筑学涉及众多学科的专业特点，因此培养优秀的建筑师是一个长线工程。当前，在中国建筑从增量发展向存量优化的进程中，我认为我们这些老同志有责任为热爱建筑设计事业的年轻人营造这种平台，培养并留住这些人才，这样我们建筑事业才会后继有人。

CA：在您的展览中我发现您工作室完成的项目涉及了各种类型，这是在市场压力下被迫做出的选择，还是您主观上更愿意尝试各类项目呢？

王振军：我们院是一个具有深厚工业背景的设计院，是这个平台让我们摒弃了对建筑类型的成见，让我们能够根据市场和项目的需求进行从容地应对。首先，在当前的信息时代的背景下和中国电子院的业务范围内，我们以"蔓·设计"理论为指导，打破了传统的"民用建筑和工业建筑二元分类法",倡导超越内容和类型的"大建筑观"。其次,我深深感谢中国电子院为我们提供了一个与世界最先进工程技术无缝衔接的平台。这使我们对建筑建构的理解更加深刻和深入,让我们有了更多实践涉及超净、抗微震、智能化和设备专业复杂系统的机会,从而让我们的建筑师得到了丰富且坚实的实践历练。

CA: 今年是中国电子院成立 70 周年,也是工作室成立的第 15 年,在这个非常关键的节点,请您谈谈对工作室的期许或展望。

王振军: 在这个具有里程碑意义的重要节点,我们举办了以"生长与蔓延"为主题的 15 年作品展。此次展览不仅展示了我们在项目类型、地域分布及作品数量等方面的持续扩展,更体现了我们对于"蔓·设计"理论的执着追求和传承。

作为建筑师,我们的工作主题将永恒地围绕思考和实践两项任务。思考,意味着从深度和广度上深入研究建筑的真谛,更深刻地理解建筑的本质;实践,则是在这种研究和理解的基础上,大胆而不懈地创作新的作品。

15 年作品展既是我们对过去实践的总结,又是我们继往开来的新起点。我们将保持对建筑的初心,坚守"蔓·设计"的理论,不断前行。

关于我们

About Us

重要作品获奖

- **上海浦东软件园一、二期工程**
  荣获信息产业部 2000、2004 年度优秀设计一等奖；
  中国建筑学会建筑创作大奖入围奖；

- **首都机场新塔台**
  荣获信息产业部 2006 年度优秀设计一等奖；

- **上海国家软件出口基地（浦东软件园三期）**
  荣获国际竞赛首奖；
  工信部 2010 年度优秀设计一等奖；

- **德国 SAP 中国研究院**
  荣获信息产业部 2008 年度优秀设计二等奖；

- **首都国际机场东区塔台**
  荣获国家优秀工程设计银质奖；
  中国建筑学会六十周年建筑创作大奖；
  工业和信息化部 2008 年度优秀设计一等奖；

- **2010 上海世博会沙特国家馆**
  竞赛荣获国际竞赛首奖；
  2010 年上海世博会创意展示 A 类金奖（国际展览局 BIE 颁）；
  第六届中国建筑学会建筑创作金奖；
  第六届中国威海国际建筑特别奖；
  2011 年度中国勘察设计协会行业优秀建筑工程设计二等奖；
  2011 年度中国勘察设计协会行业优秀智能化建筑一等奖；
  腾讯网上海世博会十大最佳展馆奖；
  2010 年工信部优秀设计一等奖；
  第十八届首都城市规划建筑设计方案汇报展优秀方案奖；

- **长沙中电软件园总部大楼**
  荣获 2013 中国建筑学会中国建筑设计奖银奖；
  北京第十七届优秀工程设计二等奖；
  荣获 2015 年度中国勘察设计协会建筑设计行业奖三等奖；

- **北京国际财源中心（IFC）**
  荣获 2013 年度中国勘察设计协会建筑设计行业奖一等奖

- **北京泰德制药股份有限公司扩建项目**
  荣获北京市 2015 优秀建筑设计二等奖

- **中国信达（合肥）灾备及后援基地项目**
  荣获 2017 年度中国勘察设计协会建筑设计行业奖二等奖；

- **中央美术学院燕郊校区教学楼项目**
  荣获 2019 年度中国勘察设计协会建筑设计行业奖一等奖；

- **北京大兴国际机场信息中心（ITC）、指挥中心（AOC）**
  荣获 2021 年度中国勘察设计协会建筑设计行业奖一等奖；

- **国家存储器基地项目（一期）**
  荣获 2021 年度中国勘察设计协会建筑设计行业奖（电子工业工程）一等奖；

- **武汉高世代薄膜晶体管液晶显示器件（TFT-LCD）生产线项目**
  荣获 2021 年度中国勘察设计协会建筑设计行业奖（电子工业工程）一等奖；

- **北京大兴国际机场生活服务设施工程**
  荣获 2021 年度电子行业优秀工程设计综合类一等奖；

- **浙江海盐智能装备制造产业基地创新中心（中欧合作产业园）**
  荣获 2021 年度电子行业优秀工程设计综合类一等奖；
  2022 年度机械工业优秀工程勘察设计项目一等奖；

重要作品获奖

- **《软件园规划设计规范》（SJ/T11448-2013）**
  荣获中国工程建设标准化协会2022"标准科技创新奖"三等奖；

- **超高世代（10.5代以上）TFT-LCD生产线工程设计关键技术研究与产业化**
  荣获中国产学研合作促进会2022年"产学研合作创新成果奖"二等奖；

- **中关村移动智能服务创新园项目**
  荣获2022年度机械工业优秀工程勘察设计项目一等奖；

Awards

- **Shanghai Pudong Software Park (Phase I / II)**
  First Prize of Ministry of Information Industry Excellent Design Award in 2000 and 2004;
  Nomination-prize of ASC Architectural Creation Award;

- **The New Control Tower for Capital Airport**
  Special Prize of Ministry of Information Industry Excellent Design Award in 2006;

- **The Conceptual Planning Design for Shanghai National Software Export Base (Pudong Software Park Phase III)**
  First Prize in International Competition;
  First Prize of the Ministry of Industry and Information Technology Excellent Design Award in 2010 ;

- **SAP China Research Institute**
  Second Prize of Ministry of Information Industry Excellent Design Award in 2008;

- **The Control Tower for East Zone, Capital Airport**
  National Excellent Engineering & Exploration Design Silver Award;
  ASC Grand Architectural Creation Award;
  First Prize of Ministry of Industry and Information Technology Excellent Design Award in 2008;

- **The Saudi Arabia Pavilion at Shanghai World Expo 2010**
  First Prize in International Competition;
  Creative Exhibition Category A Gold Prize for Shanghai World Expo 2010 (awarded by BIE);
  The Sixth ASC Architectural Creation Award;
  Special Prize of The Sixth Weihai International Architecture Award;
  Second Prize of China Engineering & Consulting Association Excellent Architectural Design Award in 2011;
  First Prize of China Engineering & Consulting Association Excellent Intelligent Building Award in 2011;
  Tencent.com Top Ten Pavilions at Shanghai World Expo 2010;
  First Prize of Ministry of Industry and Information Technology Excellent Design Award in 2010;
  The 18th Capital Urban Planning Architectural Design Schemes Exhibition Excellent Scheme Prize;

- **The Headquarters Building for Changsha CEC Software Park**
  Silver Prize of ASC China Architectural Design Award in 2013;
  Second Prize of the 17th Beijing Excellent Architectural Design Award;
  Third Prize of China Engineering & Consulting Association Excellent Architectural Design Award in 2015;

- **Beijing International Finance Center (IFC)**
  First Prize of China Engineering & Consulting Association Architectural Design Award in 2013;

- **R&D Center Expansion Project of Beijing Tide Pharmaceutical Co., Ltd.**
  Second Prize of Beijing Excellent Architectural Design in 2015;

- **China Cinda (Hefei) Disaster Recovery and Backup Base**
  Second Prize of China Engineering & Consulting Association Excellent Architectural Design Award in 2017;

- **Teaching Building Of Central Academy Offine Arts, Yanjiao Campus**
  First Prize of China Engineering & Consulting Association Excellent Architectural Design Award in 2019;

- **The ITC and AOC for Beijing Daxing International Airport**
  First Prize of China Engineering & Consulting Association Excellent Architectural Design Award in 2021;

- **National Memory Base Project (Phase I)**
  First Prize of China Engineering & Consulting Association Excellent Architectural Design Award (Electronic Industry Engineering) in 2021;

- **Wuhan High Generation Thin Film Transistor Liquid Crystal Display Device (TFT-LCD) Production Line Project**

  First Prize of China Engineering & Consulting Association Excellent Architectural Design Award (Electronic Industry Engineering) in 2021;

- **Beijing Daxing International Airport Life Service Facilities Project**

  First Prize in the Comprehensive Category of Excellent Engineering Design in the Electronics Industry in 2021;

- **Haiyan Intelligent Manufacturing Innovation Center**

  First Prize in the Comprehensive Category of Excellent Engineering Design in the Electronics Industry in 2021;
  First Prize for Excellent Engineering Survey and Design Projects in the Mechanical Industry in 2022;

- **Software Park Planning and Design Specification (SJ/T11448-2013)**

  Third Prize of the 2022 Standard Science and Technology Innovation Award of the China Engineering Construction Standardization Association;

- **Research and Industrialization of Key Technologies for Engineering Design of Ultra High Generation (10.5+) TFT-LCD Production Lines**

  Second prize of the 2022 Innovation Achievement Award for Industry University Research Cooperation of the China Association for Promoting Industry University Research Cooperation;

- **Zhongguancun Mobile Intelligent Innovation Park**

  First Prize for Excellent Engineering Survey and Design Projects in the Mechanical Industry in 2022;

王振军工作室
15 年展活动纪实
WZJ Studio
15-Year Exhibition
Documentary

（以专家现场发言顺序排序）

杨光明首席技术官主持开幕式

开幕式剪彩仪式

娄宇董事长致词

王振军总建筑师与大家分享本次展览的内容

青锋老师主持本次学术沙龙

"我们的'蔓·设计'实践"主题讲座

沙龙上半场："生长"——关于"有机建构"的几个要素

沙龙下半场："蔓延"——创作的持续性和建筑全过程把控

开幕式及学术沙龙活动大合影

王振军工作室15年展活动纪实

黄星元大师发言

庄惟敏院士发言

胡越大师发言

崔彤大师发言

邵韦平大师发言

孙宗列总建筑师发言

陈一峰总建筑师发言

薛明总建筑师发言

蔡昭昀总建筑师发言

朱铁麟总建筑师发言

蒋培铭总建筑师发言

傅绍辉总建筑师发言

董霄龙总建筑师发言

郑方教授发言

徐宗武教授发言

晁阳总建筑师发言

周栋良总建筑师发言

王振军总建筑师总结发言

王振军工作室
15年展活动纪实
15-Year Exhibition Documentary

（以专家莅临指导时间排序）

李兴钢院士、韩林飞教授莅临指导

张鹏举大师莅临指导

冯正功大师莅临指导

陈雄大师、孙一民大师莅临指导

李存东大师、《建筑学报》田华主任莅临指导

《建筑师》李鸽主编莅临指导

祁斌总建筑师莅临指导

汪恒总建筑师莅临指导

陈彬磊大师、叶依谦总建筑师莅临指导

gmp合伙人兼执行总裁吴蔚莅临指导

桂学文大师莅临指导

赵元超大师、钱方大师、总建筑师刘艺、徐宗武、吕成莅临指导

王振军工作室15年展活动纪实

杨建觉总建筑师莅临指导

马庆总建筑师莅临指导

张利大师莅临指导

活动现场

活动现场

嘉宾互动

嘉宾互动

展览现场

展览现场

展览现场

展览现场

展览现场

产、学、研及生活实录

Industry, Academia, Research and Life Record

与庄惟敏院士联合培养博士后出站

在28届UIA世界建筑师大会与崔愷院士合影

工作室门口中庭合影

口罩文创发布会

工作室年终总结座谈会

聆听大师讲座

集体讨论方案

参观大兴机场航站楼工地

新材料、新技术交流与分享

按摩医院工地现场服务

产、学、研及生活实录

大兴机场工地实录

在大兴机场生活楼屋顶

参观奇普菲尔德作品

带领北京建筑大学学生参观大兴机场行政楼工地

在出差的飞机上赶图

带着模型奔赴投标现场

出差汇报顺利通过

讨论施工图设计

参与录制《我和我的祖国》宣传片

潍坊 Mall 工地服务中

生长与蔓延　蔓·设计实践 I

一起制作手工模型

一起欢庆生日

师徒制拜师仪式

"蔓·设计"指导下的手法主义探索系列讲座

投标前的方案讨论

工作中的即时讨论

"头脑风暴"

中关村移动智能创新园项目建成后现场复盘

大兴机场行政楼工地留影

美团项目工地巡检

产、学、研及生活实录

热闹的加班餐

投标必中

与潍坊 Mall 的大中庭合影

驻场设计出征前留影

一起去郊游

驻场在烟台

建筑考察之旅

生命在于运动

记录布展时的忙碌时刻

15 年展开幕式前大扫除

生长与蔓延　蔓·设计实践 I

李天宇　　　　　　　　　　卫君

王振军

权薇　　　　　　时菲

张胜钰　　　　　　姚二将

李天宇　　　　　　卫君

产、学、研及生活实录

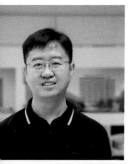 孙成伟
 李达
 陈珑
 夏璐

 徐彤
 鲍亦林
 张良钊
 郑秉东

 王舒越
 何霈
 邵宇
 许国玺

 刘敬一
 李石一
 梁赛男
 郭锦洋

 赵丽雅
 闫伟

工作室在职团队成员

主 持 建 筑 师：王振军
顾问总建筑师：张建元

成员（按入职时间排名）：

邓　涛	孙成伟	李　达	陈　珑	夏　璐	朱　谓	徐　彤
鲍亦林	张良钊	郑秉东	权　薇	时　菲	赵玮璐	王舒越
何　霈	邵　宇	许国玺	张胜钰	姚二将	丁　洋	刘敬一
李石一	梁赛男	郭锦洋	李天宇	卫　君	李　彤	赵丽雅
闫　伟						

工作室曾经团队成员

董召英	刘嘉嘉	李　晶	赵翀玺	方　雪	高　寒	王　沛
刘海傲	申　明	朱　冉	李徉贝	牟冰峰	马　蓝	胥丽娜
张宇嘉	宋熙予	张然然	……			

产、学、研及生活实录

参考文献

[1] 曾建平.自然之思——西方生态伦理思想探究[M].北京:中国社会科学出版,2004.

[2] 皮亚杰.发生认识论原理[M].北京:商务印书馆,2009.

[3] 青锋.当代建筑理论[M].北京:中国建筑工业出版社,2022.

[4] 彼得·卒姆托.思考建筑[M].张宇,译.北京:中国建筑工业出版社,2010.

[5] 史永高.材料呈现[M].南京:东南大学出版社,2008.

[6] 朱新建.打回原形[M].桂林:广西师范大学出版社,2015.

[7] 格哈德·马克.赫尔佐格与德梅隆全集[M].吴志宏译,梁蕾校.北京:中国建筑工业出版社,2010.

[8] 肯尼思·弗兰姆普敦.现代建筑 一部批判的历史[M].张钦楠等,译.北京:生活·读书·新知三联书店,2012.

[9] 肯尼思·弗兰姆普敦.建构文化研究[M].王骏阳,译.北京:中国建筑工业出版社,2007.

[10] 艾伦·G.约翰逊.见树又见林[M].喻东,金梓,译.北京:中央编译出版社,2016.

[11] 庄惟敏.建筑策划导论[M].北京:中国水利水电出版社,2000.

[12] 崔愷.本土设计[M].北京:中国建筑工业出版社,2008.

[13] 崔愷.绿色建筑设计导则[M].北京:中国建筑工业出版社,2021.

[14] 李大夏.路易·康国外著名建筑师丛书[M].北京.中国建筑工业出版社,1993.

[15] 菲利普·朱迪狄欧,珍妮特·亚当斯·斯特朗.贝聿铭全集[M].黄萌,译.北京:北京联合出版公司,2021.

[16] 大卫·E.科珀.存在主义[M].孙小玲,邓剑文,译.上海:复旦大学出版社,2012.

[17] 罗伯特·文丘里.建筑的复杂性与矛盾性[M].周卜颐,译.北京:中国建筑工业出版社,2017.

[18] 伊恩·伦诺克斯·麦克哈格.设计结合自然[M].天津:天津大学出版社,2020.

[19] 张若诗,庄惟敏.新有机建筑理念与建设性后现代主义哲学思想的关联研究"[J].世界建筑,2016 (11).

[20] 布莱恩·爱德华兹.可持续性建筑[M].周玉鹏,宋晔皓,译.北京:中国建筑工业出版社,2003.

[21] 帕·巴克.大自然如何运作[M].李炜,蔡勋,译.北京:北京大学出版社.

后　记

2017年出版了《蔓·设计》之后，我就希望将它做成为一个开放、可持续的介绍工作室和"蔓·设计"研究中心实践成果的丛书。书中实践部分系按时间轴排列，希望把工作室"蔓·设计"理论在项目中贯彻的深度和广度展现出来。

作为长期在一线工作的建筑师，繁重的生产任务指标压力之下，只能利用碎片化的时间尝试着进行系统化的思考，试图去梳理并构建自己的设计理念。让我深感欣慰的是，在"蔓·设计"理论提出到逐步完善的各个阶段，都得到了来自业内诸多专家学者和朋友们的珍贵指教。印象最为深刻的是在工作室15年展览期间，我们有机会得到国内院士、大师、老总们面对面的指教，从而使我们有机会根据他们宝贵的意见和建议对这一理论进行完善和补充。本书在把《蔓·设计》中的未尽之言及其前后的思考言论补充之后，用了尽可能多的篇幅用来展示这15年来的重要设计实践成果，是希望用直观的设计实践成果对"蔓·设计"理论所具有的指导意义做一个阶段性的验证。

感谢我们集团和设计院领导们的支持和爱护，感谢多年来一直指导我的前辈、师长和同行。

感谢黄星元大师为本书作序，并对"蔓·设计"理论给予肯定和指导意见。感谢张建元总建筑师一直在关注工作室的发展并提供了大量建设性的意见。

感谢一直关照我们的业主朋友以及项目过程中的各个专业合作团队，是我们大家的共同努力才使得项目得以在这里所呈现。

感谢清华大学的青锋老师、筑匠工程网总编张力、AT的魏星主编、UED的孙宁卿编辑、CA的牛晨曦编辑和"建筑匠人"的专题学术访谈，为读者了解我们这个团队提供了更多的视角。

感谢中国建筑工业出版社张伯熙、郑琳编辑从策划到出版全过程的严谨、专业的指导和审校，使得该书得以顺利出版。

本书的策划和编辑也花费了工作室同伴的大量心血。徐彤的组织、鲍亦林和赵玮璐的图片整理，郑秉东、丁洋的装帧设计，王舒越、何需的翻译和陈家林先生对本书提供的翻译指导；张良钊对采访文稿的审校，研究生贾一丁的专业排版，建筑摄影师周涵滔和杨超英的精彩拍摄，他们各尽其责，一起为本书最后的呈现做出了贡献。大家在一起反复讨论、推敲和比较，像对待一个项目一样层层推进的场景，将留在脑海里，成为温馨的记忆。

衷心地感谢工作室全体员工家人的共同付出，他们不仅长久以来默默地支持我们的事业，承担了全部的家务劳动，而我们常年的忙碌工作也是以缺少对他们的关心和陪伴为代价的，每每想起我内心都十分愧疚和感恩！

工作室15年的思考与实践的成果以及行业同仁们的鞭策和鼓励，使得我们更加坚定了在设计实践中坚持理论探究的信心。对建筑的热爱使我们面对挑战依旧砥砺前行，持续探索中的"蔓·设计"理念使我们对"如何看待建筑"变得更加清晰，而"如何去表达建筑"这其中的创新可能又使我们对未来的实践更加充满期待。

王振军
2024年8月31日于五路居

Epilogue

Following the publication of "Organically-Permeated Design" in 2017, I have aspired to develop it into an open, sustainable series that introduces the studio and the practical outcomes of the "Organically-Permeated Design" Research Center. The practice section of the book is arranged along a timeline, aiming to showcase the depth and breadth of the studio's implementation of "Organically-Permeated Design" theory across projects.

As an architect who has long been on the front line, the pressure of heavy production targets has meant that only fragmented time could be utilized for systematic thinking in an attempt to organize and construct my own design philosophy. It has been deeply gratifying that at every stage from the introduction to the gradual refinement of "Organically-Permeated Design" theory, I have received invaluable guidance from numerous experts, scholars, and friends within the industry. Most memorable was during the studio's 15th-anniversary exhibition, when we had the opportunity to receive face-to-face guidance from domestic academicians, masters, and chief architects, allowing us to refine and supplement this theory based on their valuable opinions and suggestions. This book, supplementing the discussions in "Organically-Permeated Design" and its preceding and subsequent thoughts, dedicates as much space as possible to showcasing significant design practice outcomes over the past fifteen years. It aims to provide a phase-wise validation of the guiding significance of the "Organically-Permeated Design" theory through tangible design practice outcomes.

I extend my gratitude to the leaders of our group company and design institute for their support and care, and to the predecessors, mentors, and peers who have guided me over the years.

I am thankful to Master Huang Xingyuan for writing the preface to this book and for affirming and guiding the "Organically-Permeated Design" theory. I appreciate Chief Architect Zhang Jianyuan for his continued attention to the studio's development and his constructive suggestions.

I am grateful to our client friends and the various professional collaboration teams involved in the projects. It is our collective effort that has made it possible to present the projects here.

I thank Professor Qing Feng from Tsinghua University, Editor-in-Chief Wei Xing from Architecture Technology (AT), Editor Sun Ningqing from Urban Environment Design (UED), Editor Niu Chenxi from Contemporary Architecture (CA), and the academic interview segment on "Architectural Craftsmen" for providing readers with additional perspectives on our team.

I am indebted to editors Zhang Boxi and Zheng Lin from the publishing house for their meticulous and professional guidance and review throughout the planning to publication process, ensuring the smooth publication of this book.

The planning and compilatiom of this book have also cost the immense dedication of our studio colleagues. Xu Tong's organization, the photo organization by Bao Yilin and Zhao Weilu, the layout design by Zheng Bingdong and Ding Yang, the translation by Wang Shuyue and He Pei, and the translation guidance provided by Mr. Chen Jialin; the review of interview manuscripts by Zhang Liangzhao, the typesetting by graduate student Jia Yiding, the excellent photography by architectural photographers Zhou Hantao and Yang Chaoying, have all contributed to the final presentation of this book. The scenes of our repeated discussions, deliberations, comparisons, and advancing step by step like a project, will remain in my mind as warm memories.

I sincerely thank the families of all the staff at the studio for their collective contributions. They have not only silently supported our endeavors for a long time, bearing the entirety of household duties, but our perennial busy work has also come at the cost of lacking in care and companionship for them, which always fills me with immense guilt and gratitude.

The outcomes of 15 years of reflection and practice of the studio, along with the encouragement and motivation from our industry colleagues, have further solidified our confidence in persisting with theoretical exploration in design practice. Our love for architecture keeps us forging ahead in the face of challenges. The ongoing exploration of the "Organically-Permeated Design" concept clarifies

后 记

our perspective on "how to perceive architecture," and the innovative possibilities in "how to express architecture" fill us with anticipation for future practice.

Wang Zhenjun
August 31, 2024, in Wuluju

- 参加集群设计——郑州航空港银 河办事处第二邻里中心小学（23地块）
- 室作品集（Ⅰ）出版
- 海崇明岛可行性研究报告完成， 相城市设计启动
- 杭州湾智能制造创新中心项目
- 中关村移动智能创新园项目
- 10年上海世博会沙特国家馆参 UIA2014 第二十五届南非德班 界建筑师大会中国建筑展"全球 程中的当代中国建筑"

- 阿克苏三馆及市民服务中心项目投标获首奖
- 北京大兴国际机场空防安保培训中心项目投标
- 潍坊崇文新农村综合服务基地概念规划及建筑方案设计项目启动
- 青岛惠普全球大数据应用研究及产业示范基地展示中心项目启动
- 北京大兴国际机场信息中心（ITC）、指挥中心（AOC）项目中标
- 中国人民银行反洗钱上海监测分析中心投标
- **北京泰德制药股份有限公司研发中心扩建项目获"2015年全国优秀工程勘察设计行业奖建筑工程三等奖"**

- 郑州阿里云中部创业创新基地项目中标
- 于海南建筑学会发表演讲"分析主义建筑学"
- **启动专著《轴线手法在当代建筑设计中的应用》编纂**
- 湖北三峡移民博物馆投标
- 塔城市文化艺术中心投标
- 北京雁栖湖山地艺术家工作区项目中标
- 中国信达（合肥）灾备及后援基地项目入选UIA2016韩国世界建筑师大会市政建筑作品展

- **工作室作品集（Ⅱ）出版**
- 北京大兴国际机场行政综合业务用房工程、生活服务设施工程、工作区物业工程、工作区车辆维修中心工程项目中标
- **专著《蔓·设计》编纂启动**
- 中国残联北京按摩医院扩建项目中标
- **参编《机场航站楼室内装饰装修工程技术规程》**
- **专著《轴线手法在当代建筑设计中的应用》出版发行**

2014　　2015　　2016

郑州航空港银河办事处第二邻里中心小学（23地块）

海盐杭州湾智能制造创新中心

北京中关村移动智能创新园

河南鄢陵·康泰半岛花语别墅区

天津逸仙科学工业园转型概念规划

阿克苏三馆及市民服务中心获首奖

北京新机场空防安保培训中心

潍坊崇文新农村综合服务基地概念规划及建筑方案设计

青岛惠普全球大数据应用研究及产业示范基地展示中心

北京大兴国际机场信息中心（ITC）&指挥中心（AOC）

中国人民银行反洗钱上海监测分析中心

北京泰德制药股份有限公司扩建项目
北京市2015优秀建筑设计二等奖

郑州阿里云中部创业创新基地

湖北三峡移民博物馆

塔城市文化艺术中心

北京雁栖湖山地艺术家工作区

北京大兴国际机场行政综合业务用房

北京大兴国际机场生活服务设施工程
2021年度电子行业优秀工程设计综合类一等奖

北京大兴国际机场工作区车辆维修中心工程

北京大兴国际机场工作区物业工程

中国残联北京按摩医院扩建项目

2017

- 北京大兴国际机场非主基地航空公司生活服务设施工程项目启动
- 河南省安阳市豫东北机场新建航站楼及塔台概念性方案项目投标
- 苏州AT论坛发表演讲"高科技园区的精神与形式"
- 专著《蔓·设计》出版发行
- 中国信达（合肥）灾备及后援基地项目获"中国勘察设计协会2017年度全国优秀工程勘察设计行业奖建筑工程设计二等奖"
- 郑州新郑国际机场二期扩建空管工程塔台小区土建及配套工程获"中国勘察设计协会2017年度全国优秀工程勘察设计行业奖优秀建筑工程设计三等奖"
- 参编《数据中心设计规范》
- 紫光南京集成电路基地项目（一期）启动

2018

- 房山区长阳镇龙湖熙悦天街项目启动
- 主编《数据中心项目规范》启动
- 北京动物园熊猫馆（冬奥馆）扩建项目启动
- 湖北鹤峰县文化体育中心项目投标启动
- 深圳中国证券期货业南方信息技术中心二期建设项目投标
- "蔓·设计"研究中心成立
- 中国电子院与清华大学联合培养谢洋博士后正式进站

2019

- 郑州中原云大数据产业园项目启动
- 参加UED"地方图式"及湖南大学建筑学科创立90周年学术沙龙谈"建筑创新"
- 烟台工程职业技术学院南校区项目中标
- 哈萨克斯坦国家灾备中心项目启动
- 烟台经济技术开发区景观卫生间项目竣工
- **获评2019中央企业劳动模范**

- 筹划工作室成立十周年活动
- "蔓·设计"标识确定
- 接受清华大学青锋教授专访
- 中央美术学院燕郊校区教学楼获"中国勘察设计协会2019年度行业优秀勘察设计优秀（公共）建筑设计一等奖"
- 接受中国勘察设计协会媒体建筑匠人专访
- 北京动物园熊猫馆（冬奥馆）扩建项目概念方案通过专家评审
- 张家口八角广场项目启动
- 字节跳动北京K项目投标

北京大兴国际机场非主基地航空公司生活服务设施

河南省安阳市豫东北机场新建航站楼及塔台概念性方案
竞赛首奖

中国信达（合肥）灾备及后援基地
中国勘察设计协会2017年度全国优秀工程勘察设计行业奖建筑工程设计二等奖

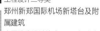
郑州新郑国际机场新塔台及附属建筑
中国勘察设计协会2017年度全国优秀工程勘察设计行业奖优秀建筑工程设计三等奖

紫光南京集成电路基地项目（一期）

芜湖皖南医学院弋矶山医院三山医养结合示范园区

房山区长阳镇龙湖熙悦天街

北京大兴国际机场非主基地航空公司生活服务设施

北京动物园大熊猫展区

湖北鹤峰县文化体育中心

深圳中国证券期货业南方信息技术中心二期
投标获首奖

潍坊龙湖产业园

郑州中原云大数据产业园

烟台工程职业技术学院南校区

哈萨克斯坦国家灾备中心

烟台经济技术开发区景观卫生间

八角湾创新产业园规划

北京大兴国际机场机场吸烟区

"蔓·设计"LOGO

中央美术学院燕郊校区教学楼
中国勘察设计协会2019年度行业优秀勘察设计优秀（公共）建筑设计一等奖

张家口八角广场

字节跳动北京K项目

北京大兴国际机场公共区派出所

北京大兴国际机场机场公安武警大基地

2020

- 烟台蓬莱国际机场信息中心可行性研究报告及概念方案完成
- 烟台中科先进材料与绿色制造山东省实验室项目中标
- 河南省汝州市临汝导航台升高改造工程建设设计项目启动
- **参编建筑学会团体标准《公共建筑后评估标准》**
- 秦皇岛博辉超高层项目方案启动
- 中国电子院与清华大学联合培养谢洋博士后出站
- 烟台祥源教育产业园启动

2021

- 与世源科技联合设计中芯京城项目
- 烟台环磁山国际科研走廊创新数字产业研究院项目中标
- 中电科四十八所半导体装备大楼项目启动
- 北京房山佛子庄乡五村更新改造概念策划项目启动
- 协同中标美团上海科技中心设计总包项目
- **海盐杭州湾智能制造创新中心获"2021年度电子行业优秀工程设计综合类一等奖"**
- **北京大兴国际机场生活服务设施工程获"2021年度电子行业优秀工程设计综合类一等奖"**

2022

- 中国同位素工程研究中心项目启动
- 中国科学院西安授时中心项目启动
- 入选《建筑实践》中国当代杰出中青年建筑师
- 承担院庆七十周年公共区域室内改造装修工作
- 潍坊云创金谷项目启动
- **专著《中外数据中心解析》编纂启动**
- 中国残联北京按摩医院扩建项目竣工
- 中国计算机博物馆投标
- 厦门天马光电子有限公司第8.6代新型显示面板生产线项目启动
- 珠海机场新建塔台及配套项目中标
- 深圳观澜导航台迁建工程及可研项目启动
- 山西大同火山云太行算力中心工程启动
- 张家口国家生物数据信息中心可行性研究报告启动

烟台蓬莱国际机场信息中心

烟台中科先进材料与绿色制造山东省实验室

河南省汝州市临汝导航台升高改造工程建设设计项目

秦皇岛博辉超高层

烟台祥源教育产业园

江西共青城吉航新材料科技有限公司厂区项目设计

中芯京城集成电路生产线

烟台环磁山国际科研走廊创新数字产业研究院

中电科四十八所半导体装备大楼

美团上海科技中心

海盐杭州湾智能制造创新中心
2021年度电子行业优秀工程设计综合类一等奖

北京大兴国际机场生活服务设施
2021年度电子行业优秀工程设计综合类一等奖

中国同位素工程

潍坊云创金谷

潍坊蓝色经济总部

山东寿光5G产业智能制造园

中央美术学院燕郊小区修建性详细规划设计

中国残联北京按摩医院扩建

计算机博物馆

厦门天马光电子有限公司第8.6代新型显示面板生产线项目

珠海机场新建塔台及配套项目

深圳观澜导航台迁建工程

山西大同火山云太行算力中心工程

- 接受《AT建筑技艺》专访

- 北京大兴国际机场信息中心（ITC）、指挥中心（AOC）工程设计获"中国勘察设计协会2021年度行业优秀勘察设计奖建筑设计一等奖"
- 湖北武汉国家存储器基地项目（一期）获"中国勘察设计协会2021年度优秀工程设计（电子工业工程）一等奖"
- 上海长电汽车芯片成品制造封测一期项目中标
- 烟台光电传感产业园项目启动
- 海尔物联网全球创新中心（三期）项目投标
- 海盐城市驿站项目启动

- 中关村移动智能服务创新园项目和海盐中欧合作产业园获"2022年度机械工业优秀工程勘察设计项目一等奖"
- 西安天和防务二期——5G通讯产业园天融大数据（西安）算力中心项目中标
- 获"光华龙腾奖/2022中国设计贡献奖银质奖"
- 北京大兴国际机场信息中心（ITC）、指挥中心（AOC）项目入选UIA2013丹麦第28届世界建筑师大会"中国建筑展"

- 工作室十五年纪念品——釉下五彩瓷笔筒出窑
- 9月19日"生长与蔓延"——王振军工作室+"蔓·设计"研究中心十五年作品展正式开幕
- 参加深圳前海紫荆项目投标
- 编纂专著《生长与蔓延——"蔓·设计"实践Ⅰ》启动

2023

- 3月19日"生长与蔓延"——王振军工作室+"蔓·设计"研究中心十五年作品展闭幕，历时6个月
- 接受UED、CA建筑专访
- 河南大学郑州校区科学展览馆、国际科技合作大厦初步设计及施工图设计项目中标
- 大连新机场空管建设工程项目中标
- 广州大湾区数字经济和生命科学产业园中标
- 字节跳动数据中心项目中标

2024

北京大兴国际机场信息中心&指挥中心
中国勘察设计协会2021年度建筑设计行业一等奖

国家存储器基地项目（一期）
中国勘察设计协会2021年度电子工业工程一等奖

上海长电汽车芯片成品制造封测项目

烟台光电传感产业园

海尔物联网全球创新中心（三期）

海盐城市驿站

中关村移动智能服务创新园
2022年度机械工业优秀工程勘察设计项目一等奖

年产10万套舰船阀门、超低温阀门及船舶氢能源阀门（易地技改）建设项目

西安天和防务二期——5G通讯产业园天融大数据（西安）算力中心

南通市北集成电路高标准厂房设计项目

东台电子产业园

釉下五彩瓷笔筒

"生长与蔓延"十五年作品展

深圳前海紫荆项目

浙江泛洋总部项目

河南大学郑州校区科学展览馆、国际科技合作大厦项目

大连新机场空管建设工程项目

广州大湾区数字经济和生命科学产业园

图书在版编目（CIP）数据

生长与蔓延：蔓·设计实践. I = GROWING AND PERMEATED ORGANICALLY-PERMEATED DESIGN PRACTICE I / 王振军著. -- 北京：中国建筑工业出版社, 2024. 12. -- ISBN 978-7-112-30774-6

Ⅰ. TU206

中国国家版本馆CIP数据核字第2024D0H898号

责任编辑：郑　琳　张伯熙
责任校对：李美娜
封面设计：郑秉东　丁　洋

生长与蔓延
蔓·设计实践 I
GROWING AND PERMEATED
ORGANICALLY-PERMEATED DESIGN PRACTICE I

王振军　著

*

中国建筑工业出版社出版、发行（北京海淀三里河路9号）
各地新华书店、建筑书店经销
北京点击世代文化传媒有限公司制版
北京雅昌艺术印刷有限公司印刷

*

开本：965毫米×1270毫米　1/16　印张：22　插页：1　字数：685千字
2024年12月第一版　2024年12月第一次印刷
定价：**328.00**元
ISBN 978-7-112-30774-6
（43885）

版权所有　翻印必究
如有内容及印装质量问题，请与本社读者服务中心联系
电话：(010) 58337283　QQ：2885381756
（地址：北京海淀三里河路9号中国建筑工业出版社604室　邮政编码：100037）